M. Gardner M.C.I.O.B., M.B.I.M.

Measurement
Level 2

Illustrated by the author

Longman Group Limited
Longman House
Burnt Mill, Harlow, Essex, UK

Published in the United States of America
by Longman Inc., New York

© Longman Group Limited 1981

First published 1981

British Library Cataloguing in Publication Data

Gardner, M
  Measurement, level 2. – (Longman technician
series: construction and civil engineering).
  1. Building – Measurement
  I. Title
  624'.028'7   TH435   80–41290

  ISBN 0–582–41584–5

Printed in Great Britain by William Clowes (Beccles)
Beccles and London

Longman Technician Series

Construction and Civil Engineering

General Editor – Construction
C. R. Bassett, B.Sc., F.C.I.O.B.
*Formerly Principal Lecturer in the Department of Building and Surveying,
Guildford County College of Technology*

Books already published in this sector of the series:

Building organisation and procedures   **G. Forster**
Construction site studies – production, administration and personnel
G. Forster
Practical construction science   **B. J. Smith**
Construction science Volume 1   **B. J. Smith**
Construction science Volume 2   **B. J. Smith**
Construction mathematics Volume 1   **M. K. Jones**
Construction mathematics Volume 2   **M. K. Jones**
Construction surveying   **G. A. Scott**
Materials and structures   **R. Whitlow**
Construction technology Volume 1   **R. Chudley**
Construction technology Volume 2   **R. Chudley**
Construction technology Volume 3   **R. Chudley**
Construction technology Volume 4   **R. Chudley**
Maintenance and adaptation of buildings   **R. Chudley**
Building services and equipment Volume 1   **F. Hall**
Building services and equipment Volume 2   **F. Hall**
Building services and equipment Volume 3   **F. Hall**
Site surveying and levelling Level 2   **H. Rawlinson**

Contents

Part 3 Estimating

Preface

This book is an introduction to Measurement and Quantity Surveying techniques and is intended primarily for building students. Two books are planned which will basically cover the syllabus of the Technician Education Council (TEC) standard units in this subject Levels II and III.

Building students enter the industry from a wide background including full- and part-time day release students and others undertaking technical courses after serving a craft apprenticeship. Few have any experience of formal measurement techniques and for most their only contact with the subject is at college. Added to this the subject is now introduced early on in a student's studies before he or she has acquired an extensive knowledge of building technology.

For these reasons I have tried to keep the book as basic as possible, seeking to lay down sound principles. Particularly in the worked examples section I have approached the taking-off on the basis of which is the simplest method to understand rather than what is the quickest method. The general aim being that at the end of two years' study the student should understand quantity surveying procedures and perhaps be able to measure quantities for a simple building.

This approach will I hope maintain interest in the subject for those students whose contact with measurement is one of just acquiring a necessary TEC unit whilst giving a firm base to those others who may go on to build a career related to this subject.

In conclusion I would like to thank colleagues at Guildford County College of Technology for their encouragement and to Mr C. R. Bassett the general editor of this series. Finally but not least to those students, who by their enthusiasm and I must admit sometimes their total lack of it for this subject, have prompted me to write this book.

Acknowledgements

Our thanks are due to the following bodies for permission to reproduce copyright material.

The Technician Education Council to quote from TEC U75/054 Measurement II. The Council reserves the right to revise the content of this unit at any time.

British Standards Institution to quote BS and BSCP reference numbers.

Standing Joint Committee for the Standard Method of Measurement of Building Works to quote references and written material from that publication.

RIBA Publications Ltd to quote from JCT Standard Form of Building Contract and reproduce an interim certificate.

RICS Publications to reproduce a valuation form.

Part One

Basic principles

Chapter One

Measurement in the building process

Introduction

The student following a course of construction studies is faced with the problem of studying measurement in association with several other subjects. In the early stages of learning measurement techniques much of the time will be spent in taking-off quantities. Because of this fact students tend to overlook the overall objectives of measurement in the construction process.

These early chapters, entitled 'Basic principles', therefore seek to place before the student the whole process of measurement, to enable the end product to be seen, although only one small part or element may be being studied at that particular time.

To facilitate this, Chapter 4 outlining a model building project from commencement to completion is included. Students should also understand that much of the book is of an introductory nature only, as some of the subject-matter would warrant a book to itself.

What is measurement

Once a client has decided to construct or alter a building and some form of drawings have been prepared outlining the scheme, there will almost certainly follow certain basic questions. These will apply whether the proposed building work is of a large or small nature and will usually include some of the following:
1. what will be the likely cost of having the work undertaken?
2. assuming an approximate cost of work can be established are any economies possible by alterations to the design and specification?
3. how should the contract for the work be placed and what type of contractors would be most suitable for undertaking the works?
4. how soon can work commence?
5. how long will the work take to complete?

To arrive at answers to these questions measurement techniques may be used. These may be fairly straightforward and simple or highly sophisticated,

4   depending on the size of contract and degree of accuracy required.

Many students new to the industry think of measurement in terms of producing bills of quantities and calculating estimates. But to the contractor the winning of a contract is just the start of an involvement in measurement processes. They must now plan the work, order materials, allocate labour and plant resources and physically carry out the work – all within the estimated contract price. To do this successfully a system of cost control must be used to compare actual construction costs against those previously estimated. This requires just as much accuracy and detail as do the more well-known forms of measurement.

Which people undertake measurement

The quantity surveyor

On projects of any size the architect will normally advise his client on the appointment of a quantity surveyor. Private quantity surveyors work in practices similarly to architects and are often called the professional quantity surveyor. They are paid a fee based on the total contract sum. Quantity surveyors are normally members of the Royal Institute of Chartered Surveyors (RICS) or of the Institute of Quantity Surveyors. They will also employ staff at technician and student level to assist with the work.

Public undertakings and local government departments have their own quantity surveying staff to control and advise on the cost of building projects.

Measurement work undertaken by quantity surveyors is varied and interesting and will include some of the following:

Cost appraisal

There are several techniques available which enable an approximate estimate of cost to be made of building projects at sketch design stage. This can assist the architect and client in establishing the most economic design for the required building work.

Production of bills of quantities

A bill of quantities is measured from the drawings and sets out to measure in detail the proposed building work in a standard manner which can subsequently be priced by contractors. The priced bill is then used throughout the contract for valuation and cost control purposes. Students are strongly advised to obtain a copy for reference purposes. Building contractors or quantity surveyors will often help out with a used one if asked.

Measurement on site

Quantities are often measured provisionally and are then subject to re-measurement on site as the actual work is undertaken. This is particularly the case with sub-structure, drainage and external works.

Valuation for stage payments

Most building contracts provide for the contractor to be paid on account as the work progresses. This is usually done on a monthly basis. The quantity surveyor will normally visit the site and evaluate the work undertaken in the current month. The valuation is then prepared in a suitable form for the architect who issues a certificate authorizing payment to the contractor.

Valuation of variations to the contract
Building contracts are usually subject to extensive variations. These are caused by a variety of reasons. The effect is to alter the design or specification and therefore the estimated price which the contractor originally submitted. The quantity surveyor should use his best endeavours to value the variations fairly to the satisfaction of the client and the contractor.

Fluctuations in cost of materials and labour
Because of the difficulty in predicting the rate of increase in the cost of materials and labour during a building project the common practice is for contractors to base their estimate at current prices. The contractor is then reimbursed as materials rise in cost or nationally agreed wage increases come into force. The contractor, therefore, does not have to predict the level of future inflation at the estimating stage.

The quantity surveyor must check and agree any claims for increased costs presented by the contractor.

Note: Some types of contracts do not allow for these fluctuations, but are usually limited to contracts not exceeding twelve months' duration.

Preparation of final account
At the end of a building project the quantity surveyor prepares the final account which is similar to a balance sheet, showing how the total contract sum has been calculated. Because of the effect of variations and fluctuations the final cost will normally exceed the estimated price and the client should be made aware of this fact.

The work of a quantity surveyor then can be likened to that of an accountant with an extensive practical knowledge of building. Remember also that the quantity surveyor is appointed to see fair play to both client and contractor.

The contractor's staff
To enable contractors to work profitably it is desirable that as many staff as possible should have a knowledge of measurement and contractual procedures. This will assist the contractor in making claims for various payments to which he is entitled under the terms of the contract.

Many companies will employ staff at a senior level who are members of the Chartered Institute of Building or Institute of Quantity Surveyors. Others may hold the national certificate or technician education certificate or have come through the trade with City and Guilds qualifications.

The contractor's surveyor: Often known as the quantity surveyor in construction companies, but to avoid confusion we will use the term 'contractor's surveyor' to distinguish this title from the previously mentioned quantity surveyor. The contractor's surveyor is employed directly by his company and his duty is entirely to his organization. Several of his functions overlap or assist those of the quantity surveyor, these include:

1. measurement on site;
2. assistance with preparation of valuations with the quantity surveyor;
3. agreeing value of variations with the quantity surveyor;
4. preparing information with regard to fluctuation claims;
5. agreeing the final account with the quantity surveyor.

In addition there are tasks which are completely independent of the quantity surveyor.

Cost reconciliation
This is to compare estimated costs with actual costs as the work progresses. Information can be fed back for use in estimating on new projects and also to advise management should a contract be going astray financially. For this type of information to be useful it must be available as quickly as possible.

Contract administration
Dealing with claims by nominated sub-contractors and suppliers and labour-only sub-contractors. Measuring and checking on site for calculation of bonus payments.

Other staff
Other members of the contractor's staff are also engaged in the measurement processes.
Estimator: The estimator must fully understand the code of measurement and how it affects the build-up of rates used for estimating purposes.
Buyer: The buying department often produces schedules of material requirements as well as negotiating the best possible prices for supply of goods.
Planners: Larger companies employ planners to identify likely building programme times. By using bulk quantities of work suitable time scales can be built up.
Contract managers: Normally have overall charge of contracts and therefore need to be well informed of financial state of contracts and of the cost control techniques employed by the company.
Site staff: Site staff should not be overlooked. They should not be overloaded with paperwork, but at the same time the correct procedures with such items as daywork sheets and material transfers are an important part of the measurement process.

Chapter Two

Some uses of measurement

Some uses of measurement

Approximate estimating.
Operational estimating.
Production of bills of quantities.
Planning and resource allocation.
Cost recording and comparison with targets.
Valuation including variations.

Introduction

This chapter sets out to indicate to students in a little depth some of the uses to which measurement is put within the construction industry. It is introductory in nature only, and before applying these techniques students would need to undertake further reading or gain practical experience in the subject.

Approximate estimating

These are techniques of cost appraisal which can be carried out by the quantity surveyor.

When a client requires building work undertaken the normal procedure is for an architect to be appointed to prepare details of suitable schemes based on the client's brief. The usual practice is for the architect to prepare sketch drawings and then consult with the client to see if the designs are broadly suitable. At this stage, as well as generally agreeing the sketch designs, the client will wish to have an indication of the likely cost of the project.

The traditional method of obtaining an estimate is for contractors to price bills of quantities which have been prepared by the quantity surveyor from the architect's working drawings. This is, however, a somewhat lengthy process. Firstly the architect must produce working drawings. These are then used by the quantity surveyor to measure a bill of quantities. Finally the completed bills are sent out to contractors for pricing. Only when these prices are received back by the architect does the client have a clear figure of the project cost. It is not unusual for this whole process to take six months and the result could be that the cost is higher than the client anticipated, or perhaps can afford. In this case a great deal of time, effort and expense would have been wasted.

It has been necessary then to develop ways of forecasting likely building costs using approximate methods at sketch design stage. This task is usually carried out by the quantity surveyor in his role as building cost consultant to

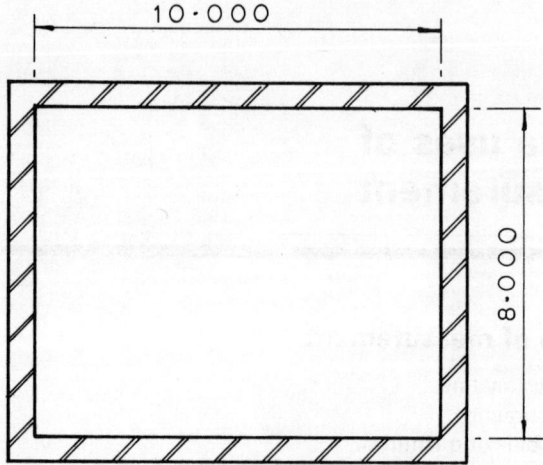

10·000

8·000

first floor plan

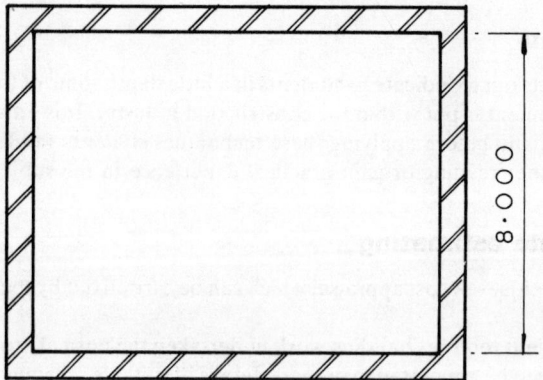

8·000

ground floor plan

ground floor plan	=	80·000 m ^{2}
first floor plan	=	80·000 m ^{2}
total	=	160·000 m ^{2}

Fig. 2.1 Superficial area method

the client. If the cost estimates are within the client's budget the whole scheme can be given the go-ahead. The traditional process is then used to establish a firm contract price.

Approximate estimating techniques are not only of use to the client. They may give valuable assistance to the architect in the comparison of sketch details, specification and building method to allow the most economic and efficient building design to evolve.

Three examples of approximate estimating techniques are now given to allow the student to understand the basis of this type of measurement:

1. the superficial area method;
2. by units of accommodation;
3. elemental unit quantities.

As with most other forms of measurement these methods range from simple straightforward applications to more sophisticated techniques which are capable of producing very accurate results. The methods used will depend on the size and complexity of the proposed building work and the degree of accuracy required in the approximate estimate.

The superficial area method

A simple method which can give reasonably accurate results if used with care and consideration. The normal practice is to calculate the area of each floor of the building within the external walls but measuring over all internal walls.

Example. 2.1 Let us assume a house has recently been built at a cost of £25,600. This is purely the building cost and does not include land, legal fees or finance charges. Figure 2.1 details the floor plans of the house.

Area of ground floor (measured within external walls) =
$$10.000 \times 8.000 = \qquad 80 \text{ m}^2$$
Area of first floor (measured within external walls) =
$$10.000 \times 8.000 = \qquad 80 \text{ m}^2$$
$$\text{total floor area of house} = \qquad \underline{160 \text{ m}^2}$$

The cost of this particular house can now be expressed in terms of price per m² to build.

$$\text{Cost per m}^2 = \frac{\text{building cost}}{\text{total floor area}} = \frac{£25,600}{160} = £160$$

Having established this cost per m² we can now use this to give an approximate estimate on future projects of a similar nature.

Example. 2.2 Six months later a further house is to be built to a similar specification and design. It is slightly larger than the previous house, measuring 10.000×10.000 internally on both floors. Using published information we ascertain that building costs have risen by 5 per cent since the estimate for the previous house was calculated. We can now calculate an approximate cost for the new property.

Area of ground floor 10.000×10.000 =	100 m²
Area of first floor 10.000×10.000 =	100 m²
total floor area of house	200 m²
Cost per m² as previously calculated	£160 m²
Add 5% increase in building cost	£ 8 m²

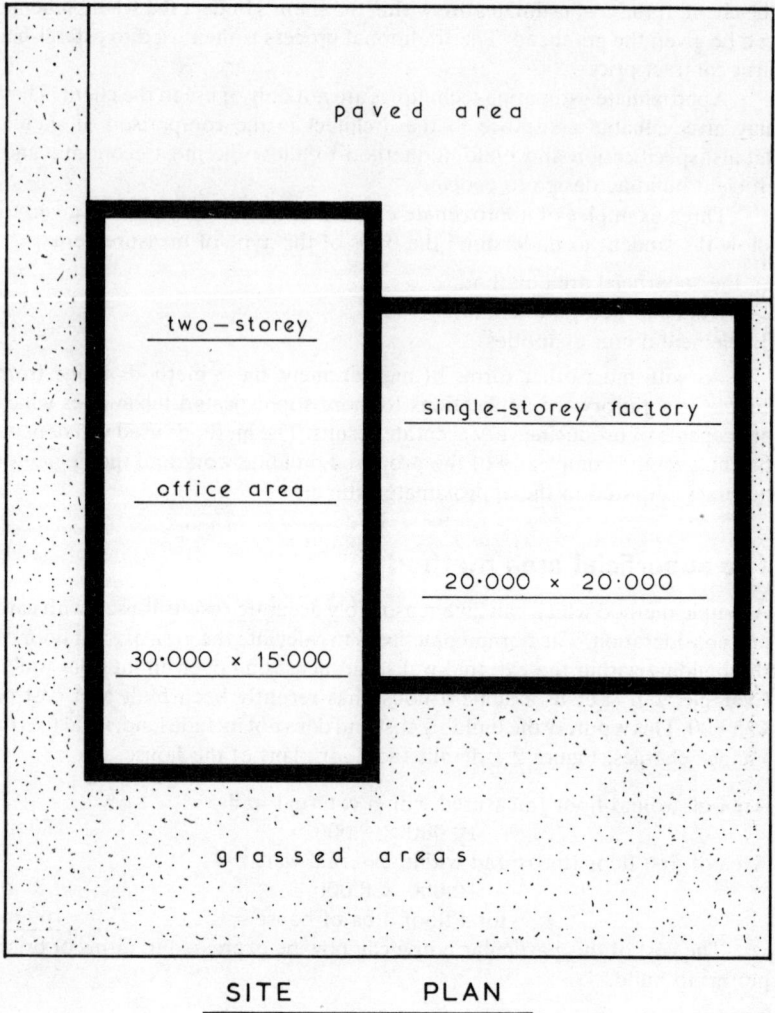

Fig. 2.2 Superficial area method

Probable cost per m² now =   £168 m²
    approximate cost for new house =
      200 m² at £168 = £33,600.
    The method may be used a stage further on larger buildings.

Example. 2.3 An industrial building is planned as detailed in Fig. 2.2. From previous contracts of a similar nature the following prices have been calculated per m² for each of the following elements.

Office areas   £180 m²
Factory areas   £160 m²
External grassed areas   £   6 m²
External paved areas   £ 10 m²

Office (two-storey)
2 × 30.000 × 15.000 =　900 m² at £180　=　£162,000
Factory (single-storey)
20.000 × 20.000　　　 =　400 m² at £160　=　£ 64,000
External grassed area
　　　　　　　　　　 =　500 m² at £　6　=　£　3,000
External paved area
　　　　　　　　　　 =　300 m² at £ 10　=　£　3,000
　　Approximate total cost　　　　　　　　=　£232,000

Limitations of the method

Students should be aware that the method gives approximate answers only:

1. buildings compared *must* be of a similar shape, specification, design and storey height;
2. contractors' estimates which are to be compared should be from similar sized companies and work capabilities.

Units of accommodation

Clients sometimes require information on building costs based on the number of persons or units of accommodation that a building is to house. The unit method endeavours to place a cost on each unit of accommodation to be provided. This may be in terms of:

1. number of beds in a hospital;
2. number of seats in a cinema;
3. number of parking spaces in a car park;
4. number of places in a school;
5. number of rooms in hotel accommodation.

To arrive at the unit price the cost of similar buildings recently erected must be studied and then expressed in terms of cost per unit.

Example. 2.4　Assume that a primary school intended to house 200 children has recently been built at a cost of £150,000.
　　The cost of the school per place is:

$$\frac{\text{building cost}}{\text{number of places provided}} = \frac{£150,000}{200} = £750 \text{ per place}$$

This method can only give very approximate answers. Type of site, specification and design could all vary the final building cost of other projects.
　　For overall budgets and calculations for investment purposes, however, it does provide a simple and quick method of approximate estimating.

Elemental unit quantities

This is a progression from other methods and will give more accurate results. The building is split up into a number of major elements which can be measured from sketch drawings to an assumed specification if necessary. These individual elements are then priced using information updated from previous contracts and are finally collected together and totalled to arrive at the approximate cost of the work.

Example of cost build-up of element

With traditional measurement techniques used in producing bills of quantities an external cavity wall would require three separate items to be measured:

1. the external wall;
2. the cavity;
3. the internal wall.

Using elemental techniques the external wall is considered as one element. A price may be built up as a whole unit by using information from priced bills of quantities from similar contracts suitably updated.

Bill of quantities description	Bill rate per m²	Add 10% increase in building costs since bill priced
External wall	£15.00	£16.50
Cavity	£ 0.50	£ 0.55
Internal wall	£ 8.50	£ 9.35

Probable cost per m² for external wall element = £26.40

Example 2.5 Figures 2.3 and 2.4 show the details of a single-storey building. In practice this would be too small to apply elemental techniques. However, it is sufficient to show the student the basis of this type of measurement. Just the structure of the building is to be costed and this may be split into three elements as indicated on Fig. 2.3 and Example 2.5a:

1. substructure;
2. external walls above d.p.c. level;
3. flat roof.

The quantities of each element may now be calculated together with the building costs updated from previous records. The quantities are measured

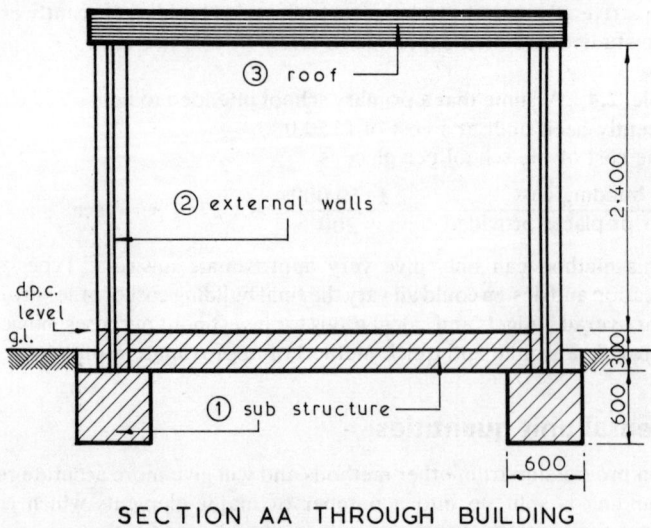

SECTION AA THROUGH BUILDING

Fig. 2.3 Elemental unit quantities

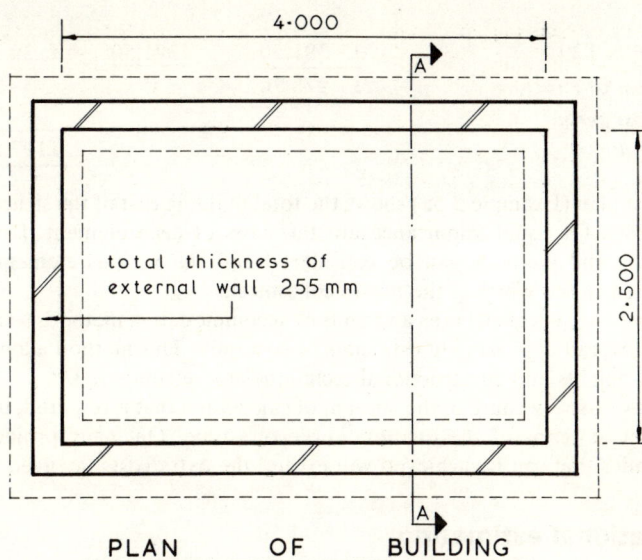

Fig. 2.4 Elemental unit quantities

using methods outlined in Chapters 5 and 6 of this book.

Example 2.5a

Element	Building cost	Quantity
Substructure		
Foundation	£40 per metre	14 m
Ground floor slab	£12 per square metre	11 m²
External walls		
Cavity construction		
consisting of facing		
bricks and blockwork	£27 per square metre	33.6 m²
Flat roof		
Timber flat roof and		
built-up felt coverings	£29 per square metre	13.5 m²

These figures may now be used in the elemental cost plan.

ELEMENTAL COST PLAN

Element		Total cost of element	Cost of element per m² of floor area (divide cost of element by floor area = 4.00 × 2.50 = 10 m²)
Substructure			
Foundation 14 m × £40 = £560			
Ground floor slab			
10 m² × £12	= £120	£ 680.00	$\frac{£680}{10}$ = £ 68.00
External wall			
33.6 m² × £27		£ 907.20	$\frac{£907.20}{10}$ = £ 90.72

Roof
    13.5 m² × £29   £   391.50 $\frac{£391.50}{10}$ = £ 39.15

    Total cost of structure   =   £1,978.70

    Total cost per m²
    of floor area   =   £197.87

The cost plan (Example 2.5a) shows the total building cost of the structure to be £1,978. Of equal importance are the costs of each element. Different materials and methods can be considered and the various element costs compared to see which is the most economical.

Both the superficial area and units of accommodation methods were able to give overall approximate estimates of cost only. This method allows plan shapes, heights and constructional techniques to be compared.

Obviously, because of the amount of calculation that is required, the cost is higher in terms of the quantity surveyor's time. On many projects the economies that can be achieved will exceed the extra costs involved.

Operational estimating

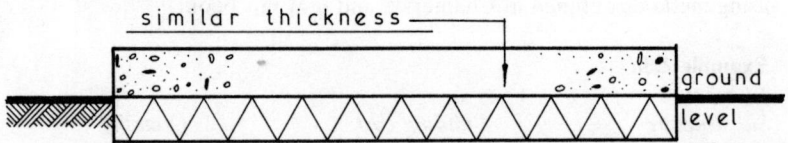

150mm thick plain concrete bed
on prepared hardcore bed of
similar thickness

ground level

SECTION   THROUGH   BASE

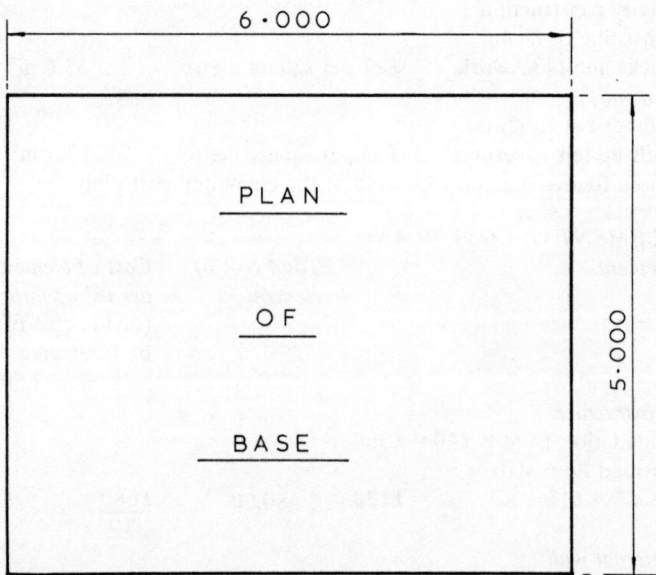

6·000

PLAN
OF
BASE

5·000

Fig. 2.5 Operational estimating

The previous types of measurement known as approximate estimating were basically calculated predictions of building costs. These are normally carried out by the quantity surveyor.

Operational estimating is invariably undertaken by the contractor's estimator and is factual in that it involves the actual calculation of building costs for the inclusion in an estimate.

In analysing building work the estimator will price under three basic headings:

1. the **material** content of the work including an allowance for waste and unloading;
2. the **labour** content of the work;
3. the **plant** content of the work.

Example 2.6 Let us assume a small contractor has been asked to submit an estimate (Example 2.6a) for the concrete base as detailed in Fig. 2.5. With small works such as this the estimator would normally measure and price his own quantities.

Example 2.6a

ESTIMATE FOR CONCRETE BASE

ESTIMATE NO. 232

MATERIALS

	£	p	£	p
1. *Concrete:* Assume ready-mixed concrete delivered to site. Volume of base 6.00 × 5.00 × 0.15 = 4.5 m³ of concrete at £27 m³	121	50		
2. *Hardcore:* Volume 6.00 × 5.00 × 0.15 = 4.5 m³ at £8 m³	36	00		
3. Compaction and wastage on hardcore 25%	9	00		
4. *Shuttering timber:* Allow 22 m. Available from our yard assumed 6 uses at 20 per m	4	40		
5. Waste on timber 10%	0	44		
6. Sundries, pegs, nails, etc.	7	00		
Total material cost	178	34	178	34

LABOUR

Assume 8 hour working day.

All-in rate £3.30 per hour.

	£	p		
1. Setting-out, excavating, wheeling away and spreading surplus excavated material on site. Allow 2 men 1 day. ∴ 8 × 2 × £3.30	52	80		

Example 2.6a (*continued*)

	£	p	£	p
2. Setting up shuttering, filling with hardcore and blind with sand (provided by client) place and tamp concrete. Allow 2 men 1 day. ∴ 8 × 2 × £3.30	52	80		
3. Striking shuttering and clearing away, leave site tidy. Allow 1 man 5 hours. ∴ 5 × £3.30	16	50		
Total labour cost	122	10	122	10
PLANT				
Pick-up truck 3 hours total delivering and collecting materials at £7 per hour inclusive of driver. ∴ 3 × 7	21	00		
Total plant cost	21	00	21	00
Total net cost			321	44
Add 20% overheads and profit			64	29
∴ estimated gross cost			385	73

Note: This example shows a very basic method of estimating. For more information the student is referred to Part 3 of this book.

The preparation of a bill of quantities

The preparation and subsequent use of bills of quantities on construction projects is the major process involving measurement techniques.

Some uses of the bill
1. to measure the work contained in the construction or alteration of building works in a systematic and standard manner in order to obtain competitive tenders from contractors for the project;
2. to assist in preparation of valuations of work in progress in order to make stage payments to the contractor;
3. to assist in the valuation of variations occurring during the progress of the project;
4. to assist the contractor in planning resources such as materials, labour and plant;
5. to assist in the preparation of the final account at the completion of the project;
6. to assist the quantity surveyor in preparing approximate estimates for future projects.

Production of the bill
The bill is prepared from the architect's drawings by the quantity surveyor in a

very exacting manner. It is measured in accordance with a code of measurement known as **The Standard Method of Measurement of Building Works**. (see Chapter 7). This code is revised periodically, the one in current use being the 6th edition commonly known as SMM6. Completed bills of quantities are the size of a large reference book.

Pricing the bill
Those contractors who are to compete for the work receive the bill of quantities with other contract documents direct from the architect.

The contractor's estimator will read through all the contract documents to get a thorough feel for the proposed work. A study of the drawings will be made. The site will be visited and a note made of any particular difficulties that may be apparent. For this purpose the estimator will often use a check list. Discussions with other members of the building team may be held to decide on the best construction methods to use. Meanwhile building material suppliers and sub-contractors' quotations will be obtained for use in pricing the bill.

Having assimilated all this information the estimator can then work through the bill calculating prices for each detailed item.

Finally the bill may be totalled and a decision made as to the percentage to be added to allow for the contractor's overheads and profit. This total sum becomes the contractor's offer of a contract price which is known as the **tender price**.

Completed tenders
Tender prices usually have to be received at the architect's office on or before a specified time and date. Tenders are then opened and scrutinized. It is the usual practice to accept the lowest tender, although the client is not bound to do so. The successful contractor's priced bills are then called for and are checked to ensure there are no arithmetical errors which may make the tender uncompetitive. Assuming that there are no such errors the client and contractor may proceed to signing the actual contract to carry out the works.

Advantages of a bill of quantities
The student will have realized that each contractor prices an identical bill. Any variance in tender prices will be due entirely to:
1. the contractor's desire to obtain the work;
2. the contractor's ability to submit a competitive estimate.

As the quantity surveyor has prepared the bill the contractor is relieved of the function of measuring quantities and can concentrate on the build-up of the estimate alone. Had each contractor to measure their own quantities there would be much duplication of effort and there is always the possibility of error and misinterpretation. The contractor is aware that any mistakes in the bill of quantities can be rectified by means of a variation and therefore the possibility of errors is reduced.

Because of the expense and time involved in the preparation of bills of quantities they are not normally used for smaller works.

Contents of a bill of quantities
A traditional bill is divided into sections under the following headings:
1. preliminaries section;
2. preamble clauses;

3. work sections;
4. provisional and prime cost sums;
5. contingency sums.

Only a brief description of each section is given below as more detailed study is undertaken at higher levels.

Preliminaries
This section, apart from giving general details of the project, includes site overhead items which cannot be included in the individual work sections. The estimator will price as a lump sum those items of relevant cost and leave unpriced those items of a general descriptive nature. Alternatively, some estimators calculate the total cost of the bill and then, based upon experience, add a percentage to cover the cost of the preliminaries. Some examples of preliminary items are:

General preliminaries:
1. description of the project;
2. description of the site;
3. form, type and conditions of contract to be used;
4. contractor's liability;
5. employer's liability.

Priced preliminaries:
1. scaffolding;
2. water for the works;
3. lighting and power for the works;
4. temporary roads, hardstandings, crossings and similar items;
5. temporary accommodation for the use of the contractor;
6. temporary fencing, hoardings, screens, fans, planked footways, guard-rails, gantries and similar items.

Preamble clauses: These may be found in two alternative positions in a bill of quantities:
1. directly after the end of the preliminaries section;
2. at the start of each work section to which they relate.

Preamble clauses provide information in a similar manner to a specification. Their purpose is to advise the estimator as to the type and standard of materials and level of workmanship expected on the project. Although the estimator does not specifically price the individual preamble clauses, as they are for communication purposes only, his subsequent pricing of the work sections may very well be influenced by the information given in them.

Work sections
The measured work is contained in work sections separated for convenience into trade headings namely:
1. demolition;
2. excavation and earthwork;
3. piling and diaphragm walling;
4. concrete work;
5. brickwork and blockwork;

6. underpinning;
7. rubble walling;
8. masonry;
9. asphalt work;
10. roofing;
11. woodwork;
12. structural steelwork;
13. metalwork;
14. plumbing and mechanical engineering installations;
15. electrical installations;
16. floor, wall and ceiling finishings;
17. glazing;
18. painting and decorating;
19. drainage;
20. fencing.

Each work section is totalled and taken to a final summary at the rear of the bill. Not all of these sections will be found in each bill, only those which are relevant to the particular project.

Provisional and prime cost sums
Provisional sum: Often when a bill of quantities is being prepared certain details of the proposed work have not been finalized. Rather than delay production of the bill of quantities an approximate sum of money is included to cover the likely cost of such work should it eventually be required. The inclusion of such a cost is known as a provisional sum. Provisional sums can be used to cover the cost of works undertaken by the main or a specialist contractor.

Prime cost sum: It is traditional for the architect to appoint specialist sub-contractors or suppliers of building materials and goods on most building projects. Such companies are known as nominated sub-contractors or nominated suppliers. An approximate sum of money is again allocated within the bill of quantities to cover each of these specialists and is known as a prime cost sum.

In practice the term 'prime cost' is often shortened to PC sum when used in the everyday context of building operations. Prime cost sums are also used in bills of quantities for work to be undertaken by statutory authorities or public undertakings such as the relevant water, gas and electricity boards.

Contingency sum
It is common practice in bills of quantities to include a lump sum of money which is known as a contingency sum. This figure is allocated to be spent entirely at the architect's discretion. Basically this allows the architect a sum of money to cover minor alterations to the project without the necessity of approaching the client for extra funds.

Note
The study of preliminaries, preamble clauses, provisional and prime cost sums is generally outside the scope of Level II work. The details that follow are for information purposes only and the student is not expected to be familiar with these parts of a bill of quantities. Students should concentrate their study on

20 the *work sections* of a bill of quantities at this level.

Figure 2.6: This shows a blank page of a bill of quantities. The column headings are shown here for the benefit of students and will not always appear in actual bills of quantities.

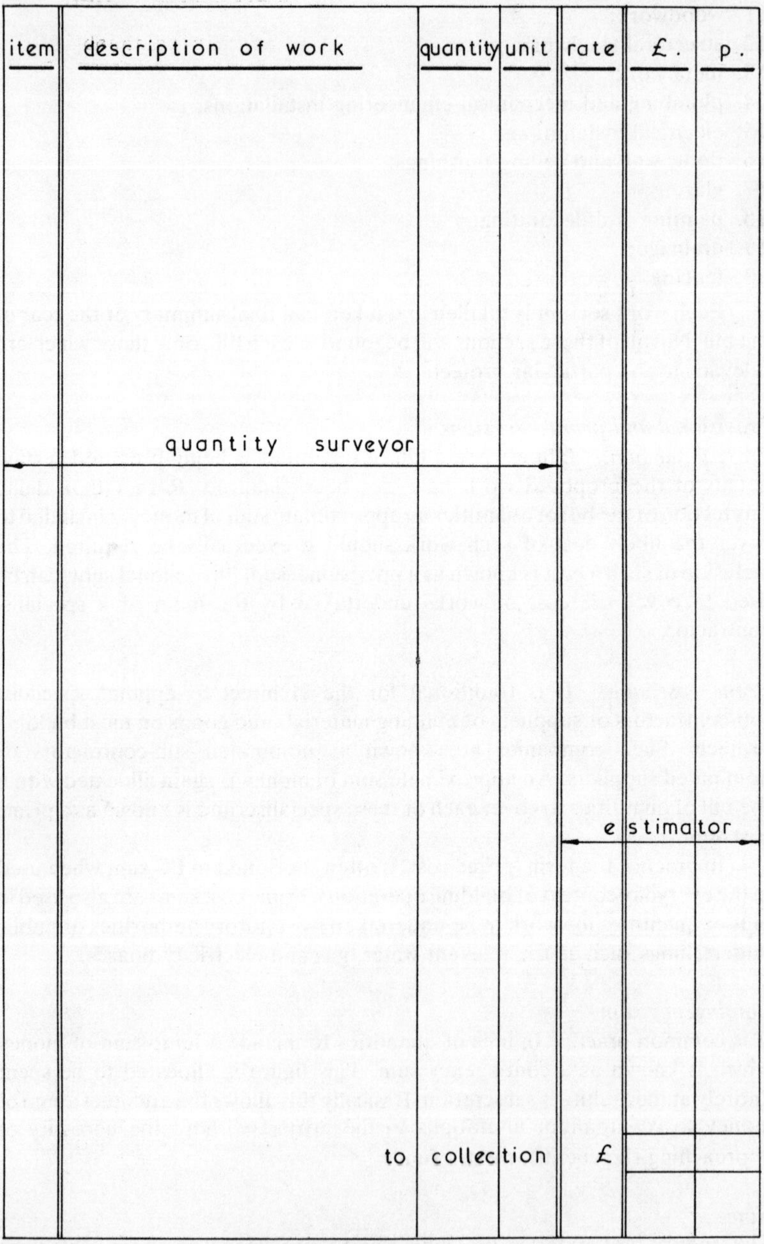

item	description of work	quantity	unit	rate	£.	p.
	quantity surveyor					
				e s tima tor		
	to collection	£				

Fig. 2.6 Blank page of bill of quantities

Figure 2.7: This is a typical page from the preliminaries section of a bill of quantities.

	BILL NO. 1					
	PRELIMINARIES AND GENERAL CONDITIONS					
A	*Employer* The Hudson Insurance Company, Landsdown Terrace, London.					
B	*Architect* James Hadmin & Partners, Consultant Architects, High View Road, London.					
C	*Quantity Surveyors* Johnson & Phillips, Chartered Quantity Surveyors, Langtree Square, London.					
	DESCRIPTION OF SITE					
D	*Location:* the site is situated in the grounds of the Hudson Insurance Company, Landsdown Terrace, London.					
E	*Access:* from Landsdown Terrace through existing car park entrance to Hudson Insurance Company.					
F	*Working area:* the contractor shall be confined to the existing car park area and alternative car parking will be provided for Hudson Insurance Company employees during the period of building operations.					
	To Collection £					

Fig. 2.7 Typical page from preliminaries section of a bill of quantities

22 *Figure 2.8:* This is a typical page of preamble clauses in a bill of quantities.

BILL NO. 2

TRADE PREAMBLES

EXCAVATION AND EARTHWORK

A   The rates included for excavation are to include for excavating in all types of ground.

B   The contractor is to give the architect 48 hours' notice when excavated trenches are ready to receive concrete. No concrete shall be deposited until the architect has inspected and approved the trench bottoms.

C   The contractor is to give the quantity surveyor 48 hours' notice when foundations are excavated to allow the measurement of any variation.

D   In the event of the contractor excavating below the given levels then the contractor will be required to fill such excavation with weak concrete (1 : 12) at his own expense.

E   The contractor shall divert as required all drains when encountered during building works. Where such works are temporary they shall subsequently be repaired to the satisfaction of the architect.

F   Filling material shall be 38 mm reject shingle.

To collection | £

Fig. 2.8 Typical page from preambles section of a bill of quantities

Figure 2.9: This is a typical work section from a bill of quantities namely 'Excavation and Earthwork'. Note the use of the various columns and the provision for totalling each page and the collection at the end of the work section. The example given clearly indicates the work of the quantity surveyor and contractor's estimator.

	BILL NO. 2					
	SUB-STRUCTURE					
	EXCAVATION AND EARTHWORK					
	Preamble clauses dealing with ground water level, nature of ground including details of trial holes and any over- or underground services present.					
	Plant					
A	The contractor is to allow for bringing to and removing from site all plant required for this section of the work.	item			120	00
B	The contractor is to allow for maintaining on site all plant required for this section of the work.	item			20	00
	Site preparation					
C	Clear site of all bushes, scrub and undergrowth including grubbing up roots.	400	m²	0·40	160	00
D	Excavate topsoil for preservation average 150 mm deep.	100	m²	0·30	30	00
	Excavation					
E	Excavate foundation trench not exceeding 0.30 m wide commencing at surface strip level average depth 0.75 m.	20	m	2·50	50	00
F	Ditto exceeding 0.30 m wide, maximum depth not exceeding 1.00 m.	24	m³	3·40	81	60
	Earthwork support					
G	Earthwork support maximum depth not exceeding 1.00 m and width between opposing faces not exceeding 2.00 m.	58	m²	0·80	46	40
		To collection		£	508	00
	50					

Fig. 2.9 Typical page from work section of a bill of quantities

					EXCAVATION AND EARTHWORK	
	Disposal of water					
A	Allow for keeping the surface of the site and the excavations free of surface water.	item			30	00
	Disposal of excavated material					
B	Deposit excavated material on site in permanent spoil heaps average 75 m from excavation.	16	m³	1·80	28	80
C	Deposit topsoil in temporary spoil heaps for re-use average 40 m from excavation.	15	m³	1·50	22	50
	Filling					
D	Filling to excavations with material arising from the excavation.	4	m³	2·00	8	00
E	Ditto with topsoil from temporary spoil heaps.	2	m³	3·00	6	00
F	Filling in making up levels average 150 mm thick with 38 mm reject shingle obtained off site.	90	m²	2·00	180	00
	Surface treatments					
G	Level and compact base of excavation to receive concrete.	130	m²	0·35	45	50
H	Blind surface of shingle filling with 25 mm sand to receive concrete.	90	m²	0·45	40	50
	Protection					
I	Allow for protection to all work in this section.	item			30	00
		To collection		£	391	30

COLLECTION				
Page	50		508	00
Page	51		391	30
	To final summary	£	899	30

51

Fig. 2.9 Work section (*continued*)

Prepares whole bill of quantities which is typewritten. This includes:

1. reference or item number column;
2. description column;
3. quantity column;
4. unit of measurement column.

Estimator

The last three columns are used by the estimator using handwritten figures:

1. the rate column;
2. multiply rate × quantity and extend answer into pounds and pence column;
3. total each page and enter into summary.

Students should realize that it is not practical within a text-book to show a whole bill of quantities. The author repeats the earlier advice to obtain a used bill of quantities from a quantity surveyor or building contractor and in conjunction with these notes become conversant with the layout and presentation.

Planning and resource allocation

Bills of quantities may be used by contractors for planning purposes once a contract has been awarded. In fact there will often appear within the preliminaries section of a bill of quantities a clause similar to that in Example 2.7.

Example 2.7

Programme and progress chart

A   The contractor shall allow for providing within three weeks from the date of possession for discussion with the architect a programme for the whole of the works including work to be carried out by nominated sub-contractors. This programme to include provisions for the recording of actual progress as the work proceeds. A copy of the programme is to be maintained and updated on site and the architect informed of any delays likely to affect completion date.

By using the bulk quantities measured in the bill of quantities the contractor can evaluate the likely time required to construct each part of the building. This information is then prepared in the form of a bar chart. An example of a simple programme is given in Fig. 2.10. Not only is the programme of assistance to the architect but it also enables the contractor to plan and make the best use of his resources. For example:

1. forward orders and delivery dates for materials can be placed based on the programme of work;
2. nominated sub-contractors can be advised well in advance of likely commencement dates of their part of the work;
3. the contractor can plan the best use of his own labour and also labour-only sub-contractors;
4. plant and equipment that will be required can be scheduled. Decisions can be made as to using the contractor's own or hired in plant.

	WEEK1	2	3	4	5	6	7	8	9	10	11	12	13
CONTRACT PROGRAMME													
sub structure	■	■											
brickwork			■	■									
roof structure					■								
roof covering						■							
first fixings							■	■					
plastering									■				
second fixings										■			
electrical										■			
decorations											■	■	
services						■	■						
drainage		■	■	■									
external work									■	■	■		
completion													■
hand over													▷

Fig. 2.10 Contract programme in bar chart form

Examples of bar chart calculation
(a) Assume that the bill of quantities shows that there are 200 m³ of concrete in the foundations. The contractor by experience knows that his labour gang can place 30 m³ per day.

$$\text{Time to place 200 m}^3 \text{ of concrete} = \frac{\text{total quantity}}{\text{daily output}} = \frac{200 \text{ m}^3}{30 \text{ m}^3}$$

$$= \underline{\text{approximately 7 days}}$$

(b) The bill shows a total area of brickwork in half-brick-wide walls of 600 m². Again from experience and past records the contractor estimates that his brickwork gang has an average output on this type of work of 4 m² per hour. Assuming a nine-hour day the daily output therefore is 4 × 9 = 36 m².

$$\text{Length of time to erect brickwork} = \frac{600 \text{ m}^2}{36} = 17 \text{ days approximately}$$

To these net allowances the contractor will add a percentage to cover delays caused by bad weather and unforeseen circumstances.

Variations and valuations

Variations to building contracts
Building works are complex in nature, and frequently variations to the original proposals are made during the progress of the work. There are several causes of variations and some of the more common are listed below:
1. errors in the preparation of bills of quantities;
2. changes to the design or specification required by the client or architect;

3. non-availability of specific materials;
4. unforeseen site conditions.

When a variation occurs for whatever reason it is the normal practice for the architect (or supervising officer on government projects) to issue an

Architect's name
and address

Works

situate at

To contractor

Under the terms of the Contract

dated

I/We issue the following instructions. Where applicable the contract sum will be adjusted in accordance with the terms of the relevant Condition.

Instructions

Architect's Instruction

Instruction no.

Date

For office use: Approx costs
£ omit   £ add

Office reference   Signed

Architect/Supervising officer

Notes

Amount of contract sum £
± Approximate value of previous instructions £
£
± Approximate value of this instruction £
Approximate adjusted total £

To Contractor ☐   Copies to Employer ☐   Quantity surveyor ☐   Clerk of works ☐   Structural consultant ☐

Heating consultant ☐   Electrical consultant ☐   ☐   ☐   Architect's file ☐

© 1967 RIBA

Fig. 2.11 Architect's instruction (reproduced by kind permission of the RIBA)

instruction in writing to the contractor informing them of the details. For convenience a pre-printed form is used known as a variation order or architect's instruction. A typical form for this purpose is detailed in Fig. 2.11. Each instruction is numbered consecutively and a copy sent to the quantity surveyor.

The student should realize that the types of contract documents used on building works vary. Variations can lead to disagreement and argument. The contractor should ensure that he follows closely the contract conditions with regard to any instruction to vary the work. Often the architect will issue a verbal instruction on site and confirm it later in writing. Many contractors will feel it prudent to keep a variation book on site so that if the architect issues a verbal instruction he can be asked to record the details and add his signature and date at that time. The use of such a book may avoid later misunderstandings.

Valuation of variations
It is the responsibility of the quantity surveyor to value the work contained in a variation. He will usually do this in consultation with the contractor's surveyor. Assuming that there is a priced bill of quantities or a schedule of rates (similar to bill but no quantities measured, just priced description of work) the quantity surveyor can use the following methods of valuing the work:

1. where the work measured in a variation is similar to that contained in the bill of quantities then the contractor's original rates may be used to calculate the cost of the variation;
2. where the work is not similar then the quantity surveyor may be able to calculate new rates which are adjusted from the contractor's original rates;
3. where the contractor's original rates cannot be used or adjusted then a fair valuation of the varied work must be agreed between the parties;
4. where none of these methods are suitable the contractor may be paid daywork. Daywork means reimbursing the contractor the direct cost of all materials, labour and plant used in the varied work plus an addition to cover the contractor's overheads and profit.

It is important to realize that not all variations will result in an additional cost. Often work is deleted and will result in an omission from the final contract cost.

Valuation of work in progress
Most building contracts make provision for the contractor to be paid on account of the work completed at regular intervals during the progress of the work. Two methods of assessing the value of work completed are:

(a) by stage payments;
(b) by interim valuations.

Stage payments: This system is commonly used for smaller contracts and housing projects although it can be used for larger works. Payment is made to the contractor as certain stages of construction are completed.

Example 2.8 A contract to extend an existing building has been awarded at an estimated price of £12,000. Payment is to be in stages as the following points in construction are reached:

Stage of building work completed	Percentage of total contract figure released	Actual payment
Sub-structure complete	20	£ 2,400.00
Plate level	35	£ 1,800.00
Roofed in	50	£ 1,800.00
Plastered	80	£ 3,600.00
Final completion	100	£ 2,400.00
Total		£12,000.00

The stage payments will normally be subject to a retention.

Interim valuations: This system is usually used for valuing work other than for smaller contracts. Interim valuations are normally carried out at monthly intervals.

The contractor's surveyor will assemble the information required which is basically an analysis of the work completed to date. Items that are included in a valuation would include the following:

1. a percentage of the total value of preliminaries (from the preliminary section of the bill of quantities);
2. the value of the contractor's work carried out to date (from the work sections of the bill of quantities);
3. the value of any variations which have been measured and their value agreed upon;
4. the value of nominated sub-contractors and nominated suppliers claims to date;
5. the value of materials on site;
6. the value of any fluctuation to date.

Checking the valuation

The quantity surveyor and contractor's surveyor will normally meet together on site to discuss the value of the work completed to date. Where sections of the bill of quantities are complete this is a fairly simple exercise. Sometimes, however, work will have to be actually measured by tape and rod to obtain an accurate valuation. Once the quantity surveyor has satisfied himself that the valuation is correct he can prepare the details in a suitable form for the architect. Standard forms are available for this purpose. The total value of work shown on a valuation is from commencement on site to when the present valuation is undertaken. To obtain the current month's valuation the total of the previous payments has to be deducted.

Example 2.9 This shows a simplified valuation laid out in a form for the student to understand. Examples of detailed valuations are given in the Joint Contracts Tribunal (JCT), *J.C.T. Guide to the Standard Form of Building Contract*, 1980 edition published by RIBA Publications Ltd.

Note: Where fluctuations are calculated using the NEDO (National Economic Development Office) formula method of price adjustment they would then be subject to retention. A summary of the valuation as shown on Fig. 2.12 may now be submitted to the architect by the quantity surveyor.

CONTRACT **NEW OFFICE BLOCK for JAMES & Co. LTD.**   VALUATION NR. **4**

MAIN CONTRACTOR **E.B. TAYLOR LTD.**   SITE DATE **2 - 6 - 80**

ISSUE DATE **3 - 6 - 80**

Valuation
£

Amounts subject to retention

Total value of:

1. Work properly executed by the contractor (includes value of work to date; proportion of total value of preliminaries; variations agreed and priced; contractor's profit on nominated sub-contractor's work) **40,000**

2. Materials and goods stored on site for incorporation in the works by the Contractor. **4,000**

3. Materials and goods stored off site for incorporation in the works by the Contractor. **—**

4. Work properly executed by nominated sub-contractors including their materials and goods stored on and off site. **20,000**

Total of amounts subject to retention **£ 64,000**

Amounts not subject to retention

5. Fluctuations payable to the contractor **300**

GROSS VALUATION **£ 64,300**

LESS

6. Retention 5% **3,200**

7. Total amount stated as due in interim certificates previously issued **45,000**   **48,200**

AMOUNT DUE **£ 16,100**

Retention

It is the normal practice to hold retention on the value of work undertaken by contractors. This serves two purposes:

(a) to encourage the contractor to complete the works as speedily as possible as half of the total retention is released upon practical completion of the building;

(b) the remainder of the retention money is released when the contractor has made good any defects during the defects liability period and the final account has been agreed.

Quantity Surveyor. of	C.Price & Partners, 12 High Street, Storton.	
Architect/S.O. of	W.H.Brown A.R.I.B.A. 110 New Road, Storton.	**Valuation**
Employer of	James & Co.Ltd. Industrial Estate, Storton.	No: 4 Date 3rd June 80
Contractor of	E.B.Taylor Ltd. Riverside Storton	QS Reference
Works at	James & Co.Ltd.	

I/We have made, under the terms of the Contract, an interim valuation as at

2nd June 1980 † and I/we report as follows:—

Gross valuation	[including nominated sub-contractors' values from attached statement *]	£	64,300·00
Less the value of any work or material notified to me/us by the Architect/S.O. in writing, as not being in accordance with the Contract		£	————
		£	64,300·00
Less retention—either [—½ of £— *]		£	
—or [from attached statement *]		£	3,200·00
		£	61,100·00
Less previously CERTIFIED		£	45,000·00
Balance (in words)	Sixteen thousand one hundred pounds .————	£	16,100·00

Contract sum   £ 150,000·00

Signature: C.Price ·   Quantity Surveyor.

Notes:
(i) All the above amounts are exclusive of V.A.T.
(ii) The balance stated is subject to any statutory deductions which the Employer may be obliged to make under the provisions of the Finance (No. 2) Act 1975 where the Employer is classed as a "contractor" for the purposes of the Act.
(iii) It is assumed that the Architect/S.O. will:—
  (a) satisfy himself that there is no further work or material which is not in accordance with the Contract.
  (b) unless otherwise agreed, notify Nominated sub-contractors of payments due to them.
  (c) satisfy himself, if he wishes, that previous payments to Nominated sub-contractors have been discharged.
(iv) The Certificate of payment should be issued within seven days of the date indicated thus†.
(v) Action by the Contractor should be taken only on the basis of figures contained in the Certificate of payment.

© 1977 RICS   * Delete as appropriate.   n.j.a.

Fig. 2.12 Quantity surveyor's valuation (reproduced by kind permission of the RICS)

Certification
Upon receipt of the quantity surveyor's valuation of current work in progress the architect issues an interim certificate to the contractor (see Fig. 2.13 for example). The contractor when he receives the certificate is entitled to payment from the client of the amount shown in accordance with the terms of the contract.

Architect's name *
and address:

W.H.Brown A.R.I.BA.
110 New Road, Storton.

**Interim
certificate**

Serial No: **0920752**

Employer's name
and address:

James & Co. Ltd.
Industrial Estate, Storton.

Issue date: **5 June 1980**
Valuation date: **3 June 1980**
Instalment No: **4**
Job reference:

Contractor's name
and address:

E B Taylor Ltd.
Riverside, Storton.

① I/We certify that in accordance with
Clause 30 of the Standard Form of Building Contract, 1963 Edition,

under the Contract

dated: **5th January 1980** in the sum of £ **150,000·00**

for the Works: **New Office Building**

situate at: **James & Co.Ltd. Industrial Estate, Storton.**

interim payment as detailed below is due from the Employer to the Contractor

Total value. . . . . **£64,300·00**
includes the value of works by nominated sub-contractors as detailed on
direction form no. **4** *dated* **3rd June 1980**

Less retention . . . . **£ 3,200·00**
after deducting any retentions released previously or herewith
(as detailed on the attached statement of retention ②)

Balance (cumulative total amount certified for payment) . . . . **£61,100·00**

Less cumulative total amount previously certified for payment . . . . . . **£45,000·00**

Amount due for payment on this certificate . . . . **£16,100·00**

(in words) **Sixteen thousand one hundred pounds**

All the above amounts are exclusive of VAT

*Issue of this form under the
name of a person who is not
registered as an architect may
constitute an offence under the
Architects Registration A. Is.*

Signed **W. H. Brown.** Architect *

Contractor's provisional assessment of total of amounts included in above
certificate on which VAT will be chargeable £ (a) %

This is not a Tax Invoice

Notes: ① Where the form of contract is the Agreement for Minor Building Works 1968, delete this line and insert
'Clause 10 of the Agreement for Minor Building Works first issued 1968'.
② Delete words in parentheses if not applicable.
③ This form may be used for the purposes of releasing retention on practical completion, on partial possesseion
or on making good defects. When used for this purpose and no statement of retention is issued, insert here
appropriate wording from the following:
'including release on practical completion/partial possession/making good defects'.

© RIBA Publications Ltd. 1977

Fig. 2.13 Interim certificate (reproduced by kind permission of the RIBA)

Payment of nominated sub-contractors and suppliers
The main contractor is responsible for making payment to the nominated
sub-contractors and suppliers from the sum received from the client.
Retention is held by the main contractor on their payments in a similar manner
to the percentage held under the main contract.
Note: The standard forms reproduced above are to be revised to accord with
the 1980 edition of the *JCT Guide to the Standard Form of Building Contract*,
but were not available at the time this book was prepared for publication.

Cost recording and comparison with targets may be grouped under a general heading of Cost Control. This is a most important factor of measurement to the contractor in that it compares actual costs with those predetermined in the estimate. Cost control techniques are carried out for several reasons:

1. to ensure that work is being carried out within the company's overall budget;
2. to compare actual site costs as the work proceeds against those calculated in the estimate;
3. to ensure that payments due to the company from various contracts are paid on time and to allow action to be taken where they are not.

The contractor's surveyor collects together the necessary records for cost control purposes. He is assisted by the accounts department who will record all

Contract Address

Contract no.

Ref.	Date		Materials	Labour	Plant
A1002	15.2.79	A.R. Jones Ltd.	100 00		
2406	18.2.79	B. Hall & Co.	200 00		
7410	19.2.79	A.B.C. Plant			100 00
	20.2.79	H. Brown Ltd.	70 00		
	20.2.79	R. Took		100 00	
	20.2.79	C. Lander		100 00	
	24.2.79	A.R. Jones Ltd.	200 00		
	26.2.79	R. Took		100 00	
	26.2.79	C. Lander		100 00	
	26.2.79	T. George		70 00	
	26.2.79	D. Harper		150 00	
	29.2.79	A.R. Jones Ltd.	300 00		
	29.2.79	B. Hall & Co.	150 00		
	29.2.79	A.B.C. Plant			100 00
		SUB-STRUCTURE COMPLETE	1020 00	620 00	200 00

TOTAL COST SUB-STRUCTURE	£1840 00
STAGE PAYMENT VALUATION	£2400 00

Fig. 2.14 Simple cash ledger

expenditure incurred on the various contracts on which the company is engaged.

Simple cost control system

A cost ledger is kept for each separate contract that the company is engaged upon. Expenditure is listed under the headings of materials, labour and plant. It is possible, then, at the end of each month to compare the value of the work undertaken with the actual cost. A summary of the financial situation of each contract can then be prepared for management showing where problems may occur.

Figure 2.14 shows a simple cash ledger in which details of cost expenditure for a particular contract are listed under the headings of material, labour and plant. Assuming that a stage payment system is being used, the cost of that section of the work is compared with the value to the contractor.

It is beyond the scope of this book to detail the finer points of cost control. The main essentials are that the records are accurate and that the necessary figures are prepared as quickly as possible to be of use. Computers are used extensively in this field as they allow quick, accurate information to be prepared and compared.

Chapter Three

Communication

Communication

In this chapter we examine some of the methods used to **communicate** information on construction projects which are part of or are related to the measurement processes.

Contract documents

These are documents that are widely used to obtain tenders for carrying out building works from contractors. Because the building industry covers such a wide variety of work ranging from minor to major works it should be apparent to the student that contract documents that are suitable for, say, a small extension project would not be adequate for larger more complex works.

The following list of contract documents details the alternative sources available, but they will not all necessarily be used on any one project. Our purpose at present is to identify the style and form of the various documents. Then in Level III we can relate them to specific types of building work.

List of contract documents

Contract documents can comprise a combination of the following:

1. drawings;
2. specifications;
3. schedules of rates;
4. schedules;
5. bills of quantities;
6. form of contract.

Drawings

Drawings are an important means of communication and are used as contract documents on all types of construction projects with the possible exception of maintenance work. Broadly speaking, we may consider drawings under three headings.

Detail design drawings
Once sketch details and preliminary schemes have been agreed with the client
work can start on the production of the detailed design drawings. Where bills
of quantities are to be prepared they are measured from the detailed design
drawings. In the course of using the drawings the quantity surveyor will usually
raise queries or seek explanation of details from the architect. During this
period, then, it is quite normal for several revisions to take place. Each revision
should be recorded because it is important that the bill of quantities accurately
reflects the work contained in the drawings. Finally the drawings reach a
degree of accuracy that enables them to be used for contract documentation
purposes.

Contract drawings
Contract drawings are the ones on which the contractor prepares his tender.
They are in fact the detailed design drawings in their finalized form. Each
drawing has a reference number, and other contract documents may refer to
them by that number. Where the conditions of contract refers to contract
drawings then it means this set.

Production drawings
The contract drawings are usually drawn to scale of 1:100 or 1:50 and serve to
give the overall details of the project. The contractor at specific points of
construction may require greater detail. Production drawings are prepared to
assist the contractor on site. They are drawn to a large scale wherever possible,
and emphasis is placed on clarity of dimensions and written details. Examples
of production drawings would include setting-out plans; foundation plans;
service layouts; details of types of frames and components used; joinery
fitments, etc.

Specifications
Specifications are usually prepared by the architect. They set out in a precise
manner a description of the work and of the standard of materials and
workmanship required on the project. As such they supplement the contract
drawings by providing additional information which is not shown on the
drawings.

On small works a bill of quantities is not viable because of the cost and
time taken to produce them being uneconomical. Contractors price the work
by referring to the drawings and specification alone and measuring the work
themselves. In this case the specification is a contract document.

Where bills of quantities are used to obtain tenders for building work any
specification details are included within the bills of quantities. It is the bill of
quantities that is considered as a contract document in this case.

Schedule of rates
We have seen that smaller works are priced using drawings and specifications
only. The contractor submits his tender in the form of a lump sum. Should a
variation occur there is no breakdown of the contractor's figures available to
allow the work to be fairly valued as there would be with a priced bill of
quantities.

Schedules of rates can be used to overcome this problem. In appearance
they are similar to a page of a bill of quantities without any quantities given.

The contractor simply inserts his rates upon which his tender is based against each description of work included on the schedule. In practice only the major work items of significant cost and value are included. Those rates can then be used to value any variations.

Schedules of rates are also used to obtain tenders for maintenance works.

Schedules

Schedules are not contract documents as such, but are sometimes included within bills of quantities or specifications. Their purpose is to communicate information in tabulated form which is easily understood. Examples are reinforcement schedules, painting and decorating schedules, ironmongery schedules, etc. It is often the practice for architects to issue schedules during the progress of a project. See also the further notes on schedules later in this chapter.

Bills of quantities

A bill of quantities provides a complete measure of the quantities of material, labour, plant and other items required to carry out a project based on the drawings, specifications and schedules. The value of the project is obtained by the contractor pricing each item.

Contract particulars

There is nothing to stop two parties, a client and a builder, drawing up their own form of contract where building work is envisaged. In practice this would be rather silly because there are many standard forms of contract available which have been revised over the years to cater for all types of building projects. These take into account all of the common problems and disputes which can occur on building projects.

Forms of contract in common use in the construction industry include those issued by the JCT (editions for both private and local government use) the form for civil engineering work issued by the Institute of Civil Engineers, whilst government work is undertaken based on Form GC/Works/1 *General Conditions of Government Contracts for Building and Civil Engineering Works.*

Documentation required at tendering, formal contract and production stages

We can now look at the way in which the contract documents previously referred to are used at the tendering, contract signing and production stages. A medium-sized project is envisaged using the JCT form of contract, private or local authorities edition, with a bill of quantities being used.

Tendering stage
1. bill of quantities;
2. drawings;
3. form of tender.

At the tendering stage the contractor receives two unpriced bills of quantities. Drawings are sufficient to show the outline and detail of the building. The full set of contract drawings may be viewed at the architect's office.

The contractor's price for the work is submitted to the architect on the form of tender.

Contract stage

The tenders are opened at the specified time and the successful tenderer (normally the one submitting the lowest tender) asked to submit his priced bill for checking, if not included with the form of tender. Assuming there are no errors found in the pricing of the bill of quantities the client and contractor can sign the contract documents which consist of:

1. the priced bill of quantities;
2. the contract drawings;
3. the form of contract.

These now constitute the contract documents and should be kept safely by the architect – or employer in the case of local authorities. The JCT form of contract is in three parts:

1. *Articles of agreement:* This is pre-printed with spaces left for names, dates and signatures.

2. *Conditions of contract:* Again pre-printed. This defines the responsibilities of the employer, contractor, architect, engineer and other non-technical matters.

3. *Appendix:* The articles of agreement and conditions of contract are separate contract documents. The appendix is part of the conditions of contract, its purpose being to supplement the conditions of contract by allowing for the insertion of information which is relevant to this specific project.

Production stage

During the progress of the work the contract documents are used in the following manner.

The contractor is required to maintain on site one copy of the contract drawing; one copy of the unpriced bill of quantities, one copy of any descriptive schedules or similar documents that relate to the project; and one copy of the master programme and any drawings that may be issued to supplement the detail contained in the contract drawings. The purpose of holding these documents on site is for reference purposes at site meetings held by the architect and to assess progress and resolve any problems. Also to ensure that the standard of workmanship and materials is to the standard specified in the various contract documents.

The priced bill of quantities will be used throughout the course of the project by the quantity surveyor and contractor for the purposes of producing interim valuations; valuing variations; for checking the bill of quantities where measurement discrepancies occur; and finally to assist in the preparation of the final account.

Note: Where small works are contemplated the bill of quantities would be replaced in the above stages by a specification and possibly a schedule of rates.

Further means of communication

Product data sheets

Product data information is produced and distributed by manufacturers of building materials and components usually on A4 size paper for ease of filing. The information falls into two categories:

1. those for sales promotion purposes to enable the product to become better known and therefore to increase sales;

2. that of a technical nature which may encourage architects to consider specifying the material for use on building projects; to quantity surveyors who may incorporate details of the product into the bill of quantities; and to contractors who may require information on using the material with regard to price, application method, handling and storage, etc.

An example of a product data sheet (Fig. 3.1) is included by kind permission of Wavin Plastics Ltd.

Installation Notes

INSPECTION CHAMBER CONSTRUCTION

The OsmaDrain Inspection Chamber is designed as an alternative to traditional construction methods for chambers of invert depths up to 910mm.

Additional excavation, deeper than that required for the drain line is not necessary and suitable backfill material as used for the drain line should be placed as a 110mm-150mm bedding to the Chamber.

Pipe connections to the Chamber are made in the same way as the standard ring seal jointing of fittings. The Blank-Off Plugs provided should be pushed into unused sockets.

Backfilling with suitable material in tamped layers should continue to the underside of the level required for a 150mm concrete plinth on which the frame of a standard rectangular Class C cover should be set. Should the top of the chamber project above the level set for the top of the plinth, it may easily be cut in-situ using a fine toothed saw.

When the Chamber is positioned in a road in situations requiring a Class B cover it will be necessary for it to be protected from traffic loadings by shuttering the external ribs of the chamber to ensure that the loadings are transferred to the surrounding ground.

When discharges into the manhole are at high velocities such as adjacent to the base of soil stacks, the Inspection Chamber Channel Covers 4D.808 left hand and 4D.809 right hand should be fitted over unused channels. The covers may be cut in half where only one channel is to be covered and are fixed by slotting the keyhole over the knob on the Blank-Off Plug in the unused socket.

RODDING

Both Chambers provide sufficient access for rodding with normal light pattern Sarawak type flexible canes, or spiral or flexible steel or coiled spring rods.

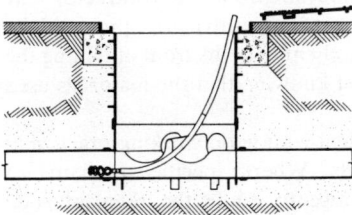

TYPICAL INSTALLATION OF 4D.857 INSPECTION CHAMBER USING CLASS C COVER

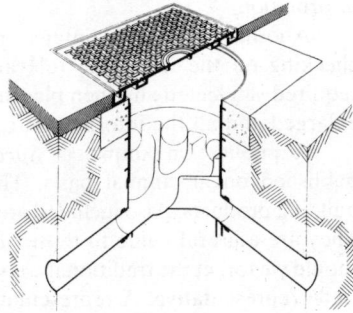

TYPICAL INSTALLATION OF 4D.891 INSPECTION CHAMBER USING CLASS B COVER

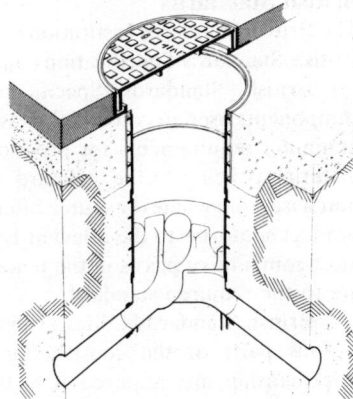

NOTE: Whilst continuing its programme of product improvement, Wavin Plastics Limited reserves the right to modify or extend any published information and to amend any product without prior notice. No responsibility can be accepted for any errors, omissions or incorrect assumptions.

Fig. 3.1   Example of a typical product data sheet

Trade catalogues

Trade catalogues are published by building material supply companies. They contain details of the products that the merchant stocks or can obtain. Trade catalogues are very comprehensive, often incorporating illustrations and specification details. Each product is given a reference number to simplify specification and ordering procedures.

Sources of information

Product data sheets can be obtained direct from the manufacturers. Trade catalogues are issued to architects, quantity surveyors and contractors' organizations who may specify or purchase materials from that company.

Many professional offices, contracting organizations and libraries maintain the international SFB/UDC filing system for product information. In the United Kingdom the code used is CI/SFB to distinguish it from the international coding.

By this stage in his studies the student should have had the system fully explained to him. Its basic concept is to provide a co-ordinated system of general product information related to the construction industry maintained within a filing system which should save much wasted time in searching for information.

Another system maintains product information on microfilm. By checking on the index the relevant microfilm containing the information required is selected and then placed in a viewer. The microfilm is then greatly enlarged and all the information can be read off.

A publication known as *Specification* by the Architectural Press, is published on an annual basis. This contains information on virtually all building products. Also included are typical specification clauses for all trades. They are a useful guide to those involved in drafting specifications. Neither should we forget the traditional conveyor of product information, the technical trade representative. A representative's primary purpose is to sell, but most are only too pleased to offer specific advice and guidance on the use of their companies' products.

British Standards

The British Standards Institution of 2, Park Street, London W1A 2BS publish British Standards Specifications and British Standards Codes of Practice.

British Standards Specifications (BSS) deal with materials and components used in construction as well as other industries, and recommend minimum requirements for products. The architect may specify the use of material as being to the standard of the relevant BSS. The contractor may purchase from whatever source offers the most competitive price provided the product conforms to the relevant BSS. The client benefits from obtaining the most competitive prices in the tender whilst knowing that the materials used are to the required standard.

British Standards Codes of Practice deal with good working practice in various parts of the construction process. Where specific standards of workmanship are required it is the practice to quote the relevant code reference number, stating that work is to be carried out in accordance with its recommendations.

Example of use of British Standards Specifications
Damp-proof courses are to be lead-cored bituminous felt to BS 743, type F.

Example of British Standards Code of Practice
Glazing: all bedding and back-puttying of glass to be in accordance with BS Code of Practice 152.

Further uses of schedules

We have seen how schedules can be included into contract documents. All members of the construction industry make use of schedules, each to a different purpose.

DOOR	SCHEDULE								
DOOR POSITION AND REF. NO.	FROM DRAWING	DOOR TYPE	DOOR SIZE	DOOR FINISH	IRONMONGERY				
					lock type	latch type	lock furniture	latch furniture	hinges
LOUNGE 1									
KITCHEN 2									
KITCHEN 3									
HALL 4									
BEDROOM 5									
BEDROOM 6									
BEDROOM 7									
BATHROOM 8									
CLOAKS 9									

Fig. 3.2 Door schedule from architect to contractor

Consider some of the uses:

Architects: To include information into the contract documents or to provide information to the contractor direct.

Quantity surveyors: Use schedules for their own internal purposes to assist in the taking-off process.

Contractors: Schedule and measure materials to obtain quotations. Compare merchants' prices in schedule form. Advise site staff of materials ordered in the form of a schedule.

SEWER SCHEDULE

drawing reference	length of excavation	average depth of excavation	length of sewer	diameter of sewer
manhole 1—2	12·00m	0·75m	12·43m	100mm ø
2—3	6·50m	1·00m	6·93m	100mmø
3—4	8·00m	1·25m	8·43m	100mmø
4—5	10·00m	1·00m	10·43m	150mmø

Fig. 3.3 Quantity surveyor's schedule

CUTTING LIST

SITE FLATS AT LION ROAD WILSHAW

MATERIAL FIRST FLOOR JOISTS

POSITION	SIZE	NUMBER	LENGTH
lounge	50×200	15	3·90
	50×200	6	3·00
	75×200	2	3·60
kitchen	50×200	10	3·00
dining room	50×200	12	3·90
hall	50×200	10	2·70
bathroom	50×200	5	2·10
bedroom	50×200	10	3·30
bedroom	50×200	11	3·00
battens	38×50	15	3·60

Fig. 3.4 Contractor's cutting list

Figure 3.2 is a simple blank schedule from architect to contractor. It depicts a door schedule for a small contract.

Figure 3.3 is a schedule as may be used by a quantity surveyor. Its purpose is to abstract information from the drawings relating to lengths, sizes and depths of excavation for a drainage system. This information is collected in schedule form before entering on to dimension paper.

Figure 3.4 is a schedule in the form of a cutting list. The materials, in this case timber, have been ordered from head office and the schedule indicates to the site staff what timber is allowed for and where. This avoids the wrong timber being used or cut and reduces site errors and waste.

Chapter Four

A model building project

Introduction

In the preceding chapters we have seen some of the basic principles of measurement. To put these into context this chapter outlines a typical medium sized building project from commencement to completion.

Assumption

It is to be assumed that the client requires a new office building of approximately £200,000 in value. Land has been acquired, finance arranged and outline planning approval obtained.

Pre-contract stage

1. Client approaches architect and outlines general requirements.
2. Architect prepares various sketch schemes for discussion with client. General agreement is reached on design.
3. Architect advises client to appoint quantity surveyor to act as cost consultant to the project. Advice is accepted and quantity surveyor is appointed accordingly.
4. The quantity surveyor prepares approximate estimates of cost based on the sketch designs. Accordingly, certain modifications are made to the proposals.
5. Architect and client agree on final details. Decision is made to invite tenders from contractors on a selective basis (see Part 3 – Estimating).
6. Architect writes to selected contractors enquiring if they are interested in tendering for the project.
7. Meanwhile work on preparing detailed drawings has carried on and planning and building by-law approval sought.
8. The local authority has now approved the project. As detailed design drawings become available they are passed to the quantity surveyor to

commence preparation of the bill of quantities. In the meantime the
architect is receiving letters back from contractors confirming their
interest in tendering for the work.

9. Production of bill of quantities has now reached draft form and is checked completely. Final printing of bill is commenced. Architect is advised.
10. Architect sends contract documents to each contractor who is to tender for the work.
11. Upon receiving information the contractor's estimator starts to involve himself in the project. Site visits are made, prices for materials obtained from suppliers, discussion on construction methods are held with management staff, possible labour and plant resources are reviewed. When all the information has been assembled pricing of the bill is undertaken. Finally, management decide on the level of profit and overheads to be added and the company's final tender figure is arrived at.
12. The completed form of tender is returned to the architect.
13. On the appointed date all tenders received are compared. Normal practice is for the lowest tender to be accepted. That contractor's priced bills are then checked for errors by the quantity surveyor.
14. Assuming there are no errors, the contract is then signed between the client and contractor.

Construction stage

15. Contractor begins to organize labour, materials and plant. Prepares programme of work and submits copy to the architect.
16. Work progresses and quantity surveyor and contractor's surveyor begin process of monthly valuations, costing variations and dealing with fluctuations claims.
17. Architect issues monthly interim certificate based on the quantity surveyor's valuation. Contractor carries out cost control checking certificate values against recorded costs of work. Payments on account are received from the client by the contractor.
18. Building is completed ready for occupation. Architect issues certificate of practical completion. Contractor is now entitled to release of half of any retention money.

Maintenance period

19. There now follows a defects liability period which is usually six months. Any faults in construction which are the responsibility of the contractor are remedied.
20. Final account is prepared. Architect issues final certificate and contractor receives balance of money outstanding, including remainder of retention monies.

Part Two

Measurement

Chapter Five

Measurement format

Introduction

This section of the book deals with the actual measurement of quantities from detailed drawings. The techniques described are traditional ones which have been practised for many years by quantity surveyors. It is fair to say that in certain cases these methods can become rather tedious and lengthy. Indeed, some students may be aware of more modern approaches which have been developed in an effort to speed up the process and allow more use of computers to be made. As with many other things in life, however, it is important to lay a solid foundation of knowledge before proceeding further. The measurement Level II syllabus is concerned with the instruction of sound basic principles in order that students may fully appreciate more advanced methods as they proceed to higher levels of study.

A further problem for some students is to reconcile formal methods of measurement used by quantity surveyors and described in this section of the book with those adopted by building contractors. Quantity surveyors use an exacting and precise system of measurement. Building companies, on the other hand, when called upon to measure their own quantities tend towards more approximate methods which rely on a combination of calculation, experience and a feel for the project in hand. Some building students question the relevance of learning such formal measurement techniques, believing them to be the province of the quantity surveyor. They should remember that later in their careers they may become contractors, surveyors or estimators. A thorough knowledge of measurement and the ways in which bills of quantities are prepared and of the relevant code of measurement will then be essential to the student.

It has also been the author's experience that the study of measurement teaches students to work with and understand construction drawings, to think deeply about the ways in which buildings are constructed and therefore to improve their knowledge of technology and to produce written and figure work in a logical clear manner. These are all desirable objectives in their own right.

Measuring quantities

The measurement processes involved in preparing a bill of quantities can be divided into four main headings namely:

1. taking-off dimensions;
2. squaring dimensions;
3. abstracting;
4. writing the bill of quantities.

Let us examine each of these stages in a little more detail.

Taking-off

This is the process of measuring quantities from detailed construction drawings. The measurements are recorded on paper known as **dimension paper**. A blank page of dimension paper is shown in Fig. 5.1. For information purposes the use to which each column is put is also shown in Fig. 5.1, although these column headings would not appear in practice. Dimension paper is divided into two equal halves and is used by writing down the left-hand side of the page from top to bottom and then similarly for the right-hand side of the page. A small binding column is often provided on the extreme left-hand edge of the paper for attaching together sheets of completed dimensions. The following points should also be noted:

1. always use the paper the correct way as Fig. 5.1;
2. number each page at the foot of the description columns for reference purposes;
3. at the head of each page write the title of contract or example being worked on.

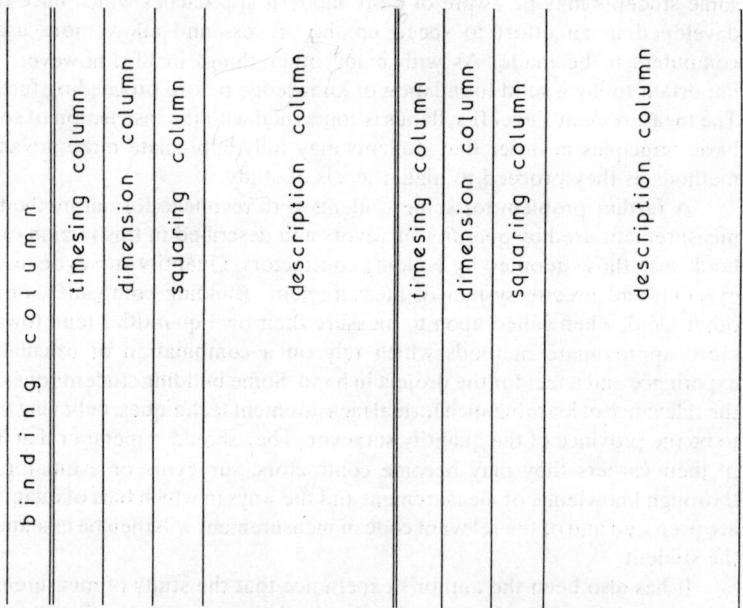

Fig. 5.1 Blank sheet of dimension paper

Squaring dimensions

The squaring up of dimensions is also carried out on dimension paper.

Squaring is done when all dimensions for a section of the work have been completed, or on smaller works, when the whole take-off is finished. As this stage involves the multiplication or adding together of figures measured initially as dimensions, an electronic calculator will be found useful.

Accuracy is important in squaring up, as indeed it is in all stages of measurement. In a quantity surveyor's office the squaring process is often undertaken by someone other than the person who actually carried out the taking-off and is carefully checked.

Examples of squaring up are given later in this chapter.

Abstracting

The preparation of the abstract follows the squaring process. The squared dimensions are transferred from dimension paper on to abstract paper. Abstract paper is double-width A3 size. The purpose of the abstract is to collect together and to arrive at total quantities for each description of work, for the subsequent inclusion into the bill of quantities.

In small examples such as contained in this book, the abstracting stage can be dispensed with. Billing can be carried out direct from the squared dimensions.

Writing the bill of quantities

Finally the bill of quantities is written up in a similar form to that already described in Chapter 2.

Use of dimension paper

Before actually measuring any work we must become familiar with the types of dimensions, the ways in which they are written down on dimension paper and also the other terms used in the taking-off process.

These may be listed as follows:

1. Types of dimensions
   (a) linear metres,
   (b) square metres,
   (c) cubic metres,
   (d) enumerated items;
2. the description;
3. bracketing;
4. squaring;
5. timesing;
6. dotting-on;
7. the ampersand;
8. correction of dimensions;
9. waste calculations.

Examples of each of these terms are now given. The descriptions are phrased from the standard method of measurement of building work which is described in Chapter 7. In these examples the right-hand half of the dimension paper is deleted in order that explanations may be given.

The dimension, description and bracketing

This simple example sets out to explain the use of these three terms. Assume that the sewer as detailed in Fig. 5.2 is to be measured between the manholes.

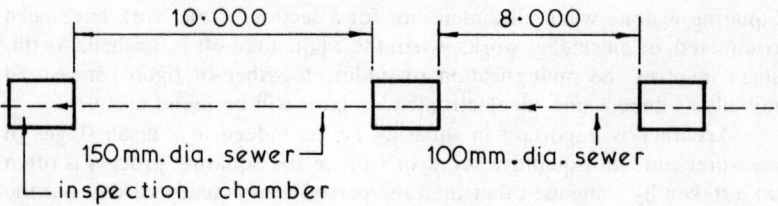

Fig. 5.2 Sewer layout

Stage 1. Calculate dimensions to two places of decimals and insert in dimension column. Drain and sewer runs are measured in linear metres.

Stage 2. Write suitable description of the work based on standard method of measurement of building works.

Stage 3. Bracket dimensions and descriptions together by drawing bracket just wide of line, separating squaring and description column.

Stage 4. Commence next dimension and description leaving space of about 40 mm.

This details the author's preferred method. Some surveyors commence with the description and then add the dimension.

In Example 5.1 the sewer is measured in linear metres. There is no need to state that the dimension is a linear one, as a single dimension with a line drawn under it indicates that the work has been measured linearly.

Example 5.1: Dimensions for Figure 5.2

8.00	100 mm diameter vitreous clay drain pipes to B.S. 65 and 540 jointed with polypropylene push fit couplings	Descriptions are usually abbreviated to save time in writing. Here they are written in full to avoid misunderstanding by the student.
		Note use of bracket to link dimensions and descriptions together.
10.00	150 mm diameter	Students usually cramp dimensions and descriptions too close together. They are far clearer to understand when spaced out.
	ditto	Aim for 40 mm between end of one item and commencement of next one.

Similarly, a square dimension is measured length × breadth or depth and would appear as follows:

10.00
3.00

Finally a cubic dimension is expressed as length × breadth × depth for example:

12.00
1.00
1.50

Examples of linear, square and cubic dimensions are given in Examples 5.2, 5.3 and 5.4.

Other terms and methods used in the taking-off process are shown in Examples 5.5 and 5.6.

Example 5.2: Linear Dimensions

<u>10.00</u>	⌉	100 mm diameter vitreous clay drain pipes to B.S. 65 and 540 jointed with polypropylene push fit couplings	This indicates a single linear dimension measured from Fig. 5.3.
<u>6.00</u>	⌉	150 mm diameter	This indicates several linear dimensions measured from Fig. 5.4. Each has the same description of work and can therefore be linked together.
<u>8.00</u>		ditto	
<u>9.00</u>			
<u>12.00</u>	⌋		

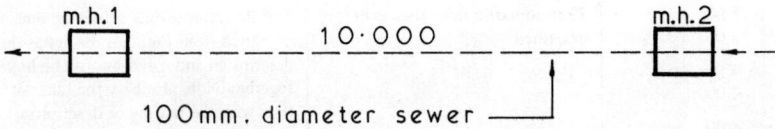

Fig. 5.3 Sewer layout

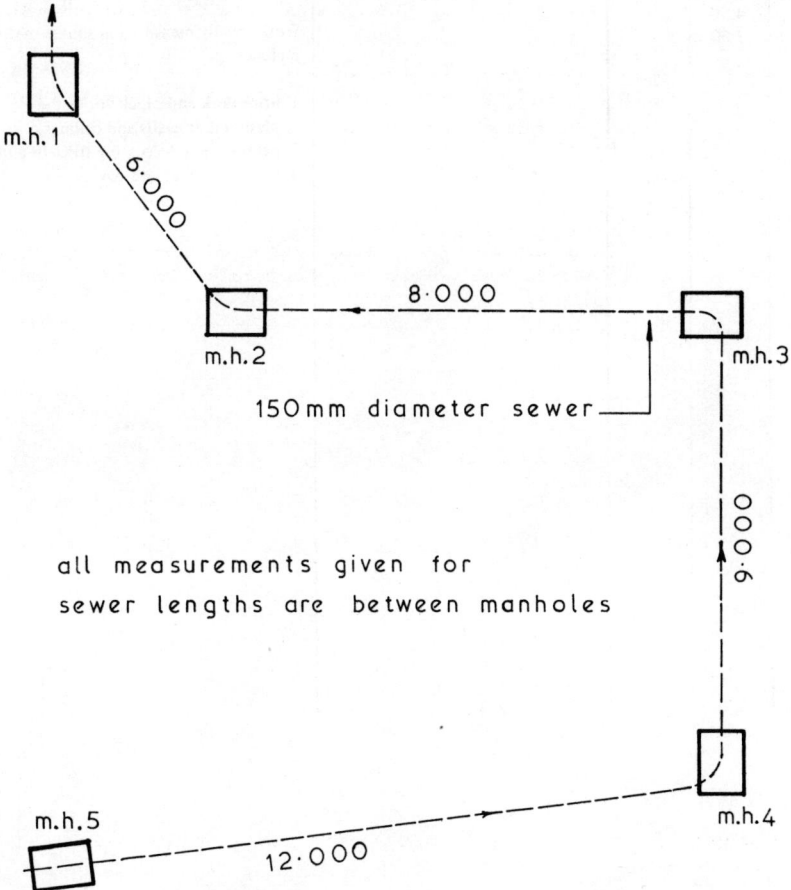

Fig. 5.4 Sewer layout

Example 5.3: Square dimensions

4.00 5.00	Thermoplastic floor tiles (P.C. £6 per m²) size 300 × 300 × 3 mm bedded in approved adhesive on level screeded bed (measured separately)

This indicates a single square dimension measured from Fig. 5.5. Square dimensions are expressed as

length × breadth

or

length × depth

5.00 3.00 4.50 4.00 4.00 3.50 4.50 3.00	Thermoplastic floor tiles as before described

This indicates several square dimensions measured from Fig. 5.6. Each has the same description and can therefore be linked together. Note also how the taker-off can save time with the writing of descriptions where the work is of a similar nature to that previously measured.

Work usually measured in square metres includes:

1. brickwork and blockwork;
2. plastering to walls and ceilings;
3. painting and decorating to walls and ceilings;
4. floor finishings and coverings.

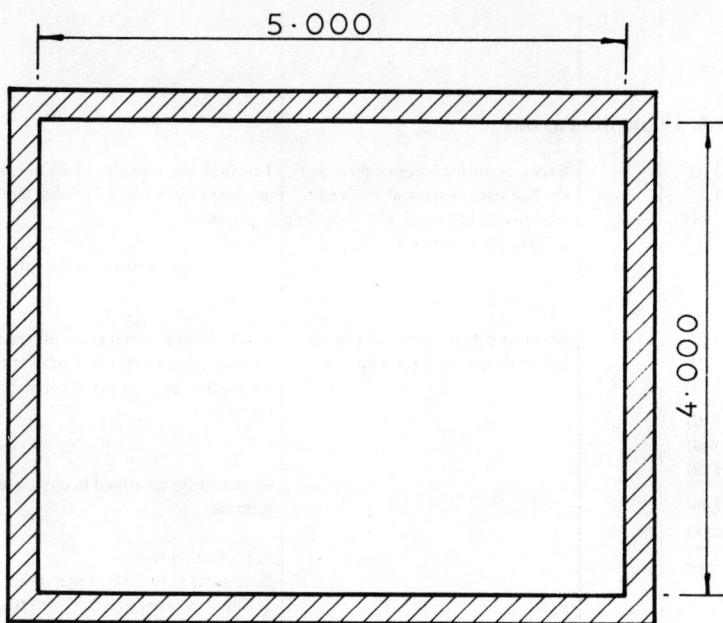

Fig. 5.5 Plan of single room

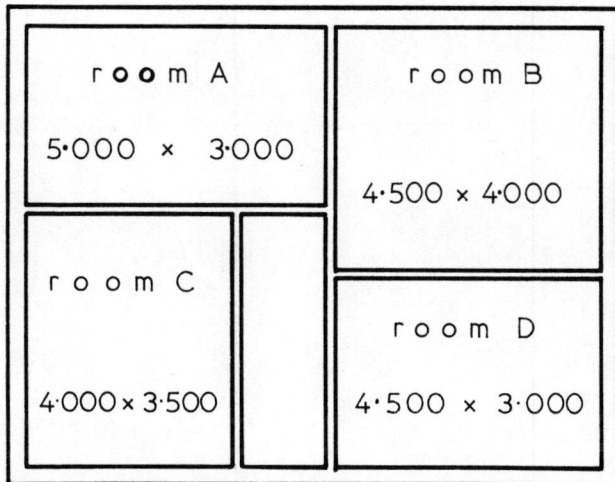

Fig. 5.6 Plan of building

Example 5.4: Cubic dimensions

1.25 1.25 <u>1.10</u>	Excavate pit for bases of piers and the like commencing at reduced level maximum depth not exceeding 2.00 m (in 1 number)	This indicates a single cubic dimension measured from Fig. 5.7. Cubic dimensions are expressed as: length × breadth × depth
1.25 1.25 <u>1.40</u> 1.40 1.40 <u>1.50</u> 1.60 1.60 <u>1.75</u>	Excavate pits for bases of piers as before described (in 3 number)	This indicates several cubic dimensions measured from Fig. 5.8. Each has the same description and can therefore be linked together. Work usually measured in cubic metres includes: 1. excavation work; 2. disposal of excavated material; 3. filling material over 250 mm thick; 4. concrete work.

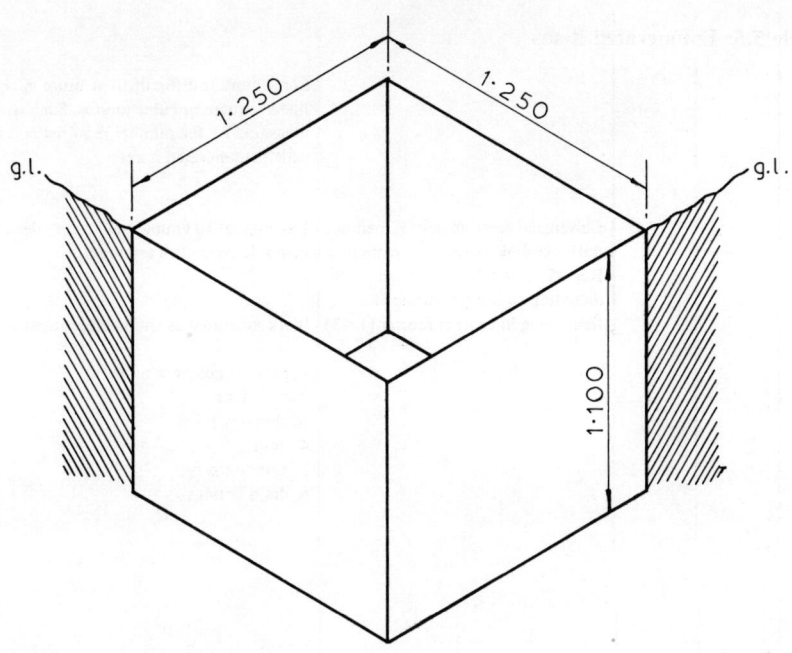

Fig. 5.7 Isometric details of foundation base

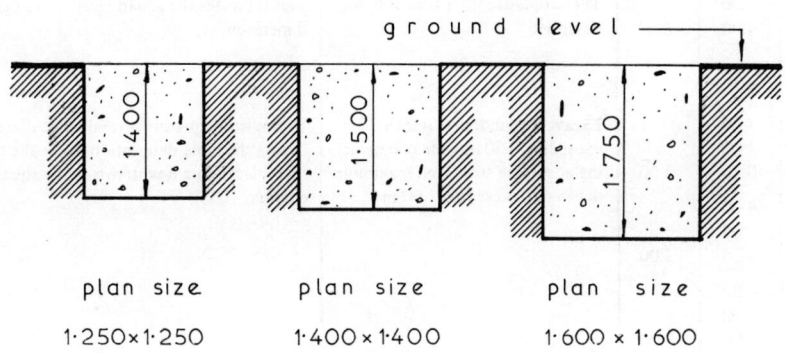

Fig. 5.8 Sections through foundation bases

Example 5.5: Enumerated items

Some work is difficult to measure in terms of linear, square or cubic metres. Such work is measured by the number required and are called enumerated items.

2		Galvanized steel single-seal medium duty manhole cover and frame to B.S. 497 size 600 X 450 mm including bedding pointing and flaunching in cement mortar (1 : 3)

Example of an enumerated item, meaning two manhole covers are required.

Work measured as enumerated items includes:

1. pre-cast concrete units;
2. air bricks;
3. chimney pots;
4. doors;
5. ironmongery;
6. drain fittings.

Example 5.6: Squaring dimensions

Squaring is the process of multiplying out each set of dimensions to arrive at the total quantity The answer is then placed opposite the squared dimension in the squaring column.

2.00		Thermoplastic floor tiles as before described
4.00	8.00	

This indicates the squaring up of a single dimension.

7.00		Excavate foundation trench exceeding 0.30 m wide commencing at surface strip level maximum depth not exceeding 1.00 m
1.00		
0.50	3.50	
4.00		
1.00		
0.75	3.00	
8.00		
1.00		
1.00	8.00	
	14.50	

Example of squaring several cubic dimensions having the same description. Note the total is then placed in a box drawn in the squaring column.

2.00		Thermoplastic floor tiles as before described
3.00	6.00	
4.00		
5.00	20.00	
7.00		
3.00	21.00	
	47.00 m^2	

An alternative method is to total the squared dimensions under a line drawn at the bottom of the description.

Many surveyors prefer to square figures in a different colour to make them more legible.

Often in building work dimensions repeat themselves. For example Fig. 5.9 shows a plan of a building. It can be clearly seen that three of the rooms are 3.00 m × 3.00 m in size and the other four are 4.00 m × 3.00 m. If we were measuring the floor areas the calculations would be as follows:

ROOMS A, B and C $3.00 \times 3.00 \times 3 = 27$ m²
rooms D, E, F and G $4.00 \times 3.00 \times 4 = 48$ m²

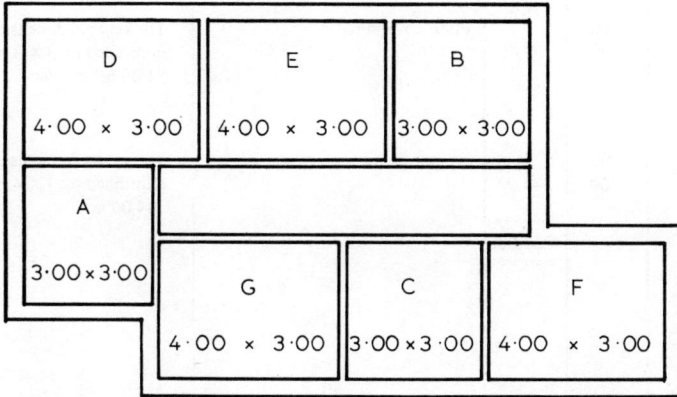

Fig. 5.9 Plan of building

The same dimensions can be set down on dimension paper by making use of the timesing column. The figures in the dimension column are simply multiplied by those in the timesing column. To write the above calculation on dimension paper we should proceed as shown in Example 5.7.

Example 5.7: Timesing

3/	3.00 3.00	27.00	Floor coverings	This indicates that the squared dimension of 9.00 is to be multiplied by the 3 in the timesing column. The squared total is 27.00 m
4/	4.00 3.00	48.00		This indicates that the squared dimension of 12.00 is to be multiplied by 4 making a total quantity of 48.00 m².
		75.00		

Timesing can be taken a stage further. Let us assume that the same building is to be three storeys high and that each floor layout is identical. By writing a further figure 3 in the timesing column followed by an oblique line we can indicate that the total number of rooms is 3 × 3 = 9. This is shown in Example 5.8.

Example 5.8				
$^3/_3/$	3.00 3.00	81.00	Floor coverings	The squared dimension of 9.00 is multiplied by 3 × 3 making a total of 81.00 m².
$^3/_4/$	4.00 3.00	144.00		The squared dimension of 12.00 is multiplied by 3 × 4 making a total of 144.00 m².
		255.00		

When using the timesing column the taker-off should try to represent the dimensions as closely as possible to those on the drawings. Hence timesing 3 × 3 gives a clearer indication of the number of rooms and storeys than simply timesing by 9.

Although square dimensions are given here as examples the timesing process can be applied in the same way to linear or cubic dimensions and to enumerated items.

Dotting on

When timesing dimensions the taker-off will on occasions need to add. Again the timesing column is used, but this time a dot is used to indicate addition – see Example 5.9.

The ampersand (&)

When measuring work certain descriptions will have the same-sized dimensions. The taker-off may use the ampersand sign to link the descriptions to the same dimension to save writing the dimension again. The ampersand sign is commonly known as 'anding on' and is written in the description column as & – see Example 5.10. Several descriptions can be 'anded on' in this manner as Example 5.11 clearly illustrates.

Example 5.9: Dotting on

2˙	3/	3.00		Floor coverings
		3.00	45.00	

This indicates that the squared dimension of 9.00 is to be timesed by 3 making a total of 27.00. To this must be added 2 × 9.00 making a total of 45.00 m².

1˙	3/	4.00	
		3.00	48.00

This indicates that the squared dimension of 12.00 is to be timesed by 3 plus 1.

	93.00

Example 5.10: The ampersand

25.00
2.40

Render and set to blockwork walls

&

Two coats of emulsion paint to plastered walls

Example 5.11

20.00
10.00

Hardcore filling average 150 mm thick in making up levels

&

Level and compact ground under

&

Blind surface of hardcore with sand

62 Correction to dimensions

Errors will be made in taking-off quantities as with any other type of work. Where alterations have to be made to dimensions it is better to remove the original dimension and write it out again in as neat a manner as possible – see Example 5.12.

Example 5.12

~~10.00~~ ~~2.00~~ 10.00 2.50	NIL	Hardcore filling in making up levels average 150 mm thick	The incorrect dimension has the word 'NIL' written in the squaring column beside it. The dimension may also be lined through if required.

Waste calculations

Working drawings often do not show the specific dimensions that the taker-off requires. The taker-off must then resort to calculation to find the desired dimensions. Such calculations are called waste calculations. The following points should be observed when writing waste calculations:

1. they should be written as far as possible in the description column of the dimension paper but may spread over the other columns when necessary;
2. they must be set down clearly and logically. Mental arithmetic should not be used and each stage should be capable of checking for accuracy;
3. waste calculations are normally written preceding the description to which they refer. Sometimes, however, it is convenient to calculate several of the waste calculations required at the commencement of taking-off.

An example of waste calculation is given in Example 5.13.

Order of taking-off

This should follow the order of the *Standard Method of Measurement of Building Works*. Generally speaking the measurement can be divided into distinct sections namely:

1. the structure;
2. finishings and services.

Example 5.13: Waste calculation

Length of building	10.000	Note the use of double underline to indicate the dimension required at the end of the waste calculation.
Add 2/205	0.410	
Length of excavation	10.410	
Length of building	7.055	Where the third digit behind the decimal point is 5 or more the second figure is rounded up by 1 when transferring to the dimension column.
Add 2/205	0.410	
Length of excavation	7.465	

10.41	Excavate topsoil for preservation
7.47	average 150 mm deep.

The structure
Foundation work; external and internal walls; floors; roofs.

Finishings and services
Wall finishings; adjustment of openings for windows and doorways; plumbing installation; drainage; external works.

Conclusion
To those students who are not familiar with taking-off methods there may appear an awful lot to be learnt. When working through the exercises later in this book they should quickly become conversant with methods explained in this chapter.

Chapter Six

Mensuration

Introduction

In this chapter methods outlined in Chapter 5 are applied to simple mensurational problems including use of:

1. geometric area and volume formulae;
2. gradients;
3. compensation lines;
4. Pythagoras' theorem;
5. elementary trigonometrical ratios.

The same methods are applied to common mensurational problems found in building measurement, namely:

6. measurement overall and adjustment of wants and voids;
7. centre-lines and corner adjustments;
8. interpolation of levels.

Geometric area and volume formulae

In Chapter 5 we studied the basic techniques and terms used in the taking-off process. When measuring we often make use of geometric area and volume formulae. Figures 6.1, 6.2, 6.3 and 6.4 show some of the more common area and volume shapes found in simple building work. The student should first study the shape and the formula and then note the manner in which it is written on to dimension paper. It is important that the correct sequence of measurement as explained in Chapter 5 is followed:

linear : length
square : length × breadth or height
cubic : length × breadth × depth

SQUARE OR RECTANGLE

5·00
4·00   Area

5·000
a
b   4·000

Area = base × height = ab

TRIANGLE

$\frac{1}{2}$ | 5·00
      4·00   Area

b
a
5·000
4·000

Area = $\frac{1}{2}$ base×height = $\frac{1}{2}$ ab

CIRCLE

π | 2·50
     2·50   Area

2·500
r

Area = πr^2

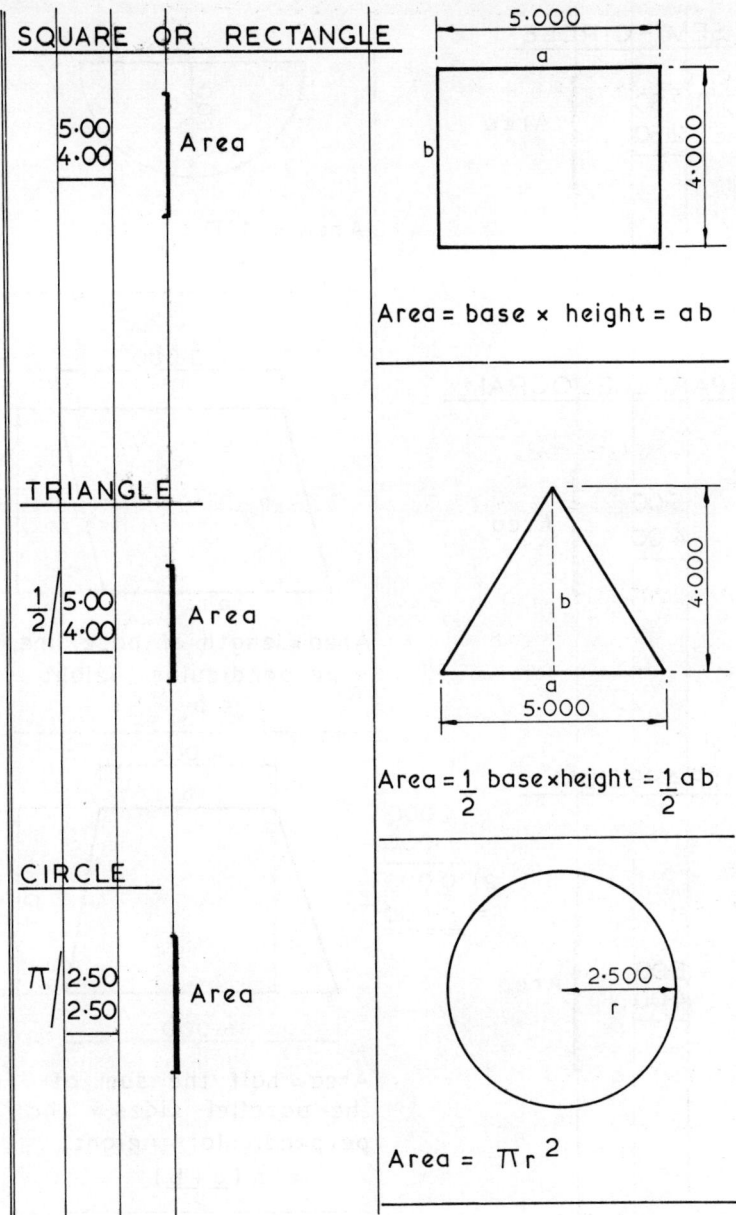

Fig. 6.1 Area and volume formulae expressed in dimension format

66

SEMI-CIRCLE

$\frac{1}{2}\left/\pi\right/\begin{matrix}2\cdot50\\2\cdot50\end{matrix}$ Area

2·500 r

$$\text{Area} = \frac{1}{2}\pi r^2$$

PARALLELOGRAM

$\begin{matrix}5\cdot00\\4\cdot00\end{matrix}$ Area

5·000

4·000

b

a

Area = length of base line
× perpendicular height
= a b

TRAPEZIUM

$\begin{matrix}4\cdot000\\6\cdot000\\\hline 2)10\cdot000\\= \underline{5\cdot000}\end{matrix}$

$\begin{matrix}5\cdot00\\4\cdot00\end{matrix}$ Area

4·000

b

h

a

4·000

6·000

Area = half the sum of
the parallel sides × the
perpendicular height
$$= h\frac{(a+b)}{2}$$

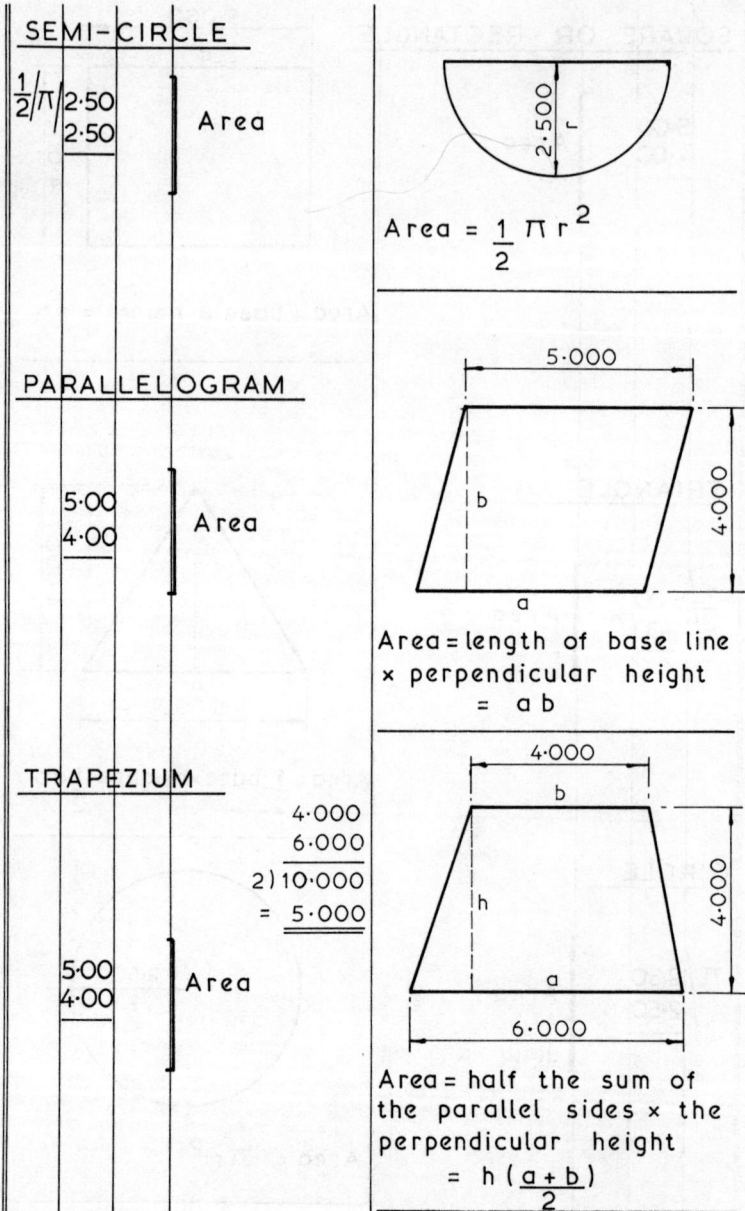

Fig. 6.2 Area and volume formulae expressed in dimension format

COMPOSITE SHAPE

$\frac{1}{2}$ / 4·00
2·00

4·00
3·00

$\frac{1}{2}$ / π / 2·00
2·00

Area

RECTANGULAR PRISM

2·00
3·00
4·00

Volume

Volume = area of end section × length = a b l

CYLINDER

π / 1·25
1·25
5·50

Volume

Volume = area of end section × length = $\pi r^2 l$

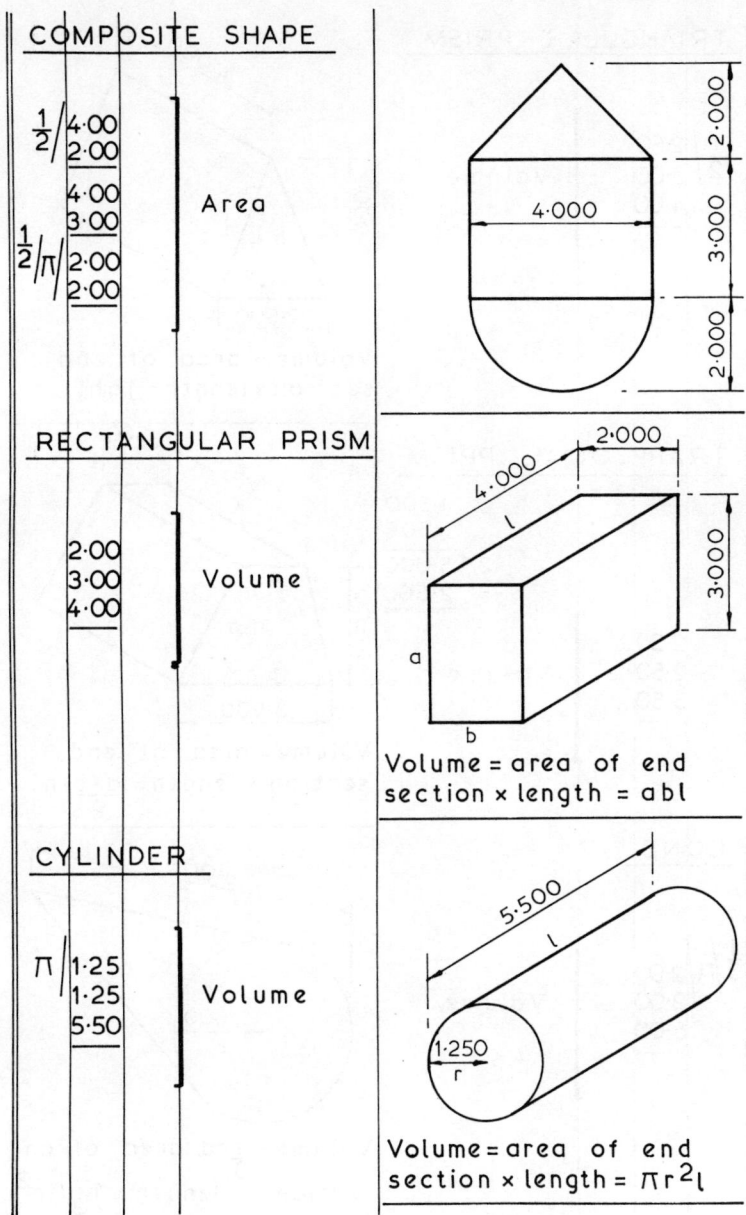

Fig. 6.3 Area and volume formulae expressed in dimension format

TRIANGULAR PRISM

$\frac{1}{2}$ | 2·50
3·00
4·00

Volume

Volume = area of end
section × length = $\frac{1}{2}$ahl

TRAPEZOIDAL PRISM

1·500
3·500
2) 5·000
= 2·500

2·50
2·50
3·50

Volume

Volume = area of end
section × length = $\frac{(a+b)}{2}hl$

CONE

$\frac{1}{3}$ | π | 2·00
2·00
6·00

Volume

Volume = $\frac{1}{3}$rd (area of end
section × length) = $\frac{1}{3}\pi r^2 l$

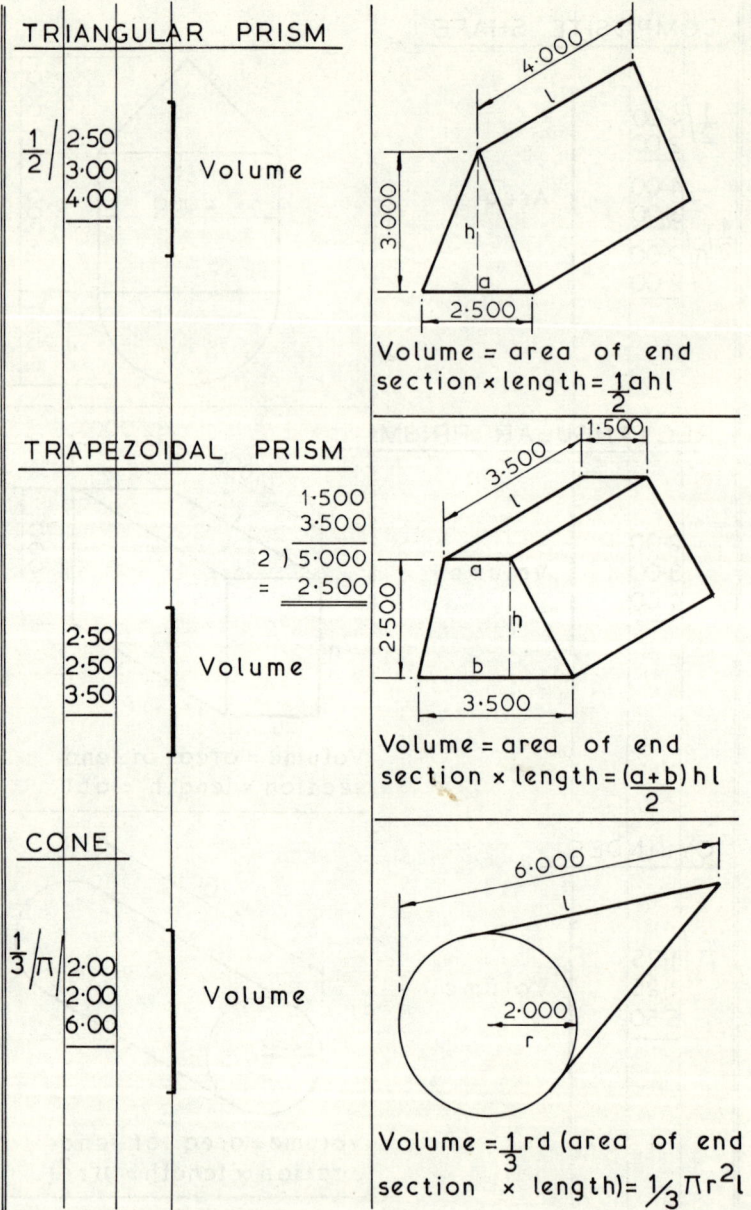

Fig. 6.4 Area and volume formulae expressed in dimension format

Gradients

A required surface fall or slope in building work may be specified as a gradient.

A gradient of 1 in 10 simply means that there is a vertical rise or fall of 1 unit in 69 a horizontal length of 10 units (see Fig. 6.5). In certain measurement problems it may be necessary to calculate the length of the surface slope given the

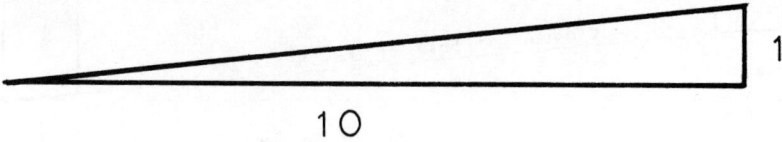

Fig. 6.5 Gradient of 1 in 10

gradient required. Sometimes we may have to incorporate our knowledge of gradients into problems involving the calculation of depths of sewers and the like. Two examples of the use of gradient calculation in measurement are now given.

Example 6.1 Figure 6.6 details a section through an embankment to the side of a motorway. Assuming that the sloping surface c is to be seeded then we need to calculate that length in order to measure the area.

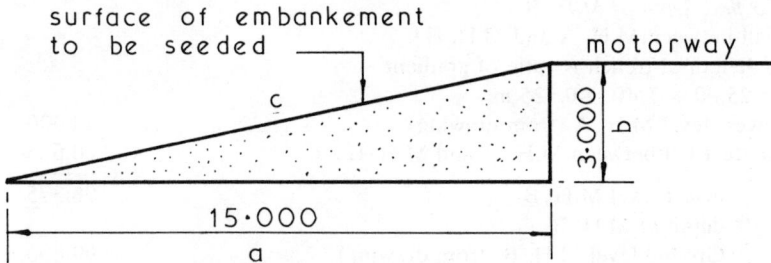

Fig. 6.6 Section through embankment

The required gradient is 1 in 5. The height of the motorway b is 3.00 m and as the gradient is 1 in 5 the base line a of the triangle so formed must be 15.00 m. Probably the simplest way, then, to calculate length c is to use Pythagoras.

$$\text{length } c = \sqrt{(a^2 + b^2)}$$
$$\text{length } c = \sqrt{(225 + 9)}$$
$$\text{length } c = \sqrt{(234)}$$
$$\text{length } c = \underline{15.30}$$

The length of surface slope for seeding is therefore 15.30 m.

Example 6.2. Figure 6.7 shows a detail of a sewer excavation for a drain which is to be laid at a gradient of 1 in 40. For measurement purposes we need to know the following:

1. The actual depths of all manholes (M.H.). In this case the invert level of M.H. A which is existing is 99.000. Therefore, we need to calculate the depth of M.H. B only.
2. Sewer trench excavation is measured in linear metres stating the average depth. We require, therefore, to calculate the average depth of trench excavation between M.H. A and M.H. B.

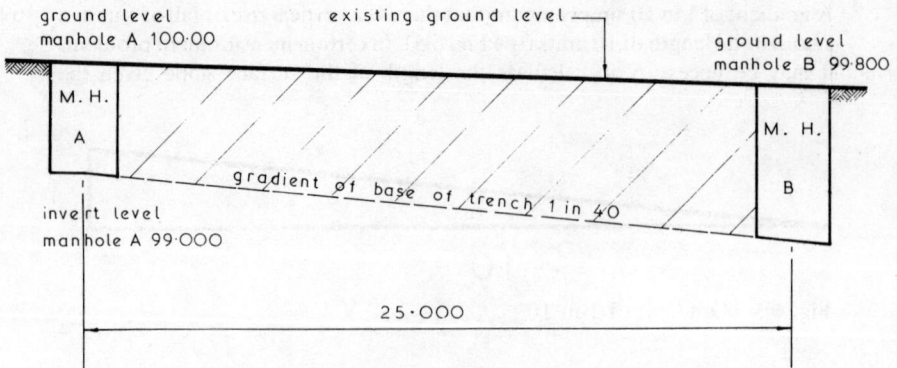

ground level
manhole A 100·00   existing ground level

ground level
manhole B 99·800

M. H.

A

gradient of base of trench 1 in 40

M. H.

B

invert level
manhole A 99·000

25·000

drawing is diagramatic only calculate depth of manhole B
and average depth of excavation.

Fig. 6.7 Sewer gradients

We may proceed as follows. The depth of M.H. A and the gradient of the
excavation is given.

To find depth of M.H. B.
Fall between M.H. A and M.H. B
= length of trench × ratio of gradient
= 25.00 × 1/40 = 0.625 m

Invert level M.H. A (from drawing)		99.000
Deduct fall between M.H. A and M.H. B	=	0.625
invert level M.H. B	=	98.375
depth of M.H. B		
Ground level M.H. B (from drawing)		99.800
Invert level M.H. B		98.375
Depth of M.H. B	=	1.425 m

Having calculated the depth of M.H. B it is now simple to find the average
depth of trench between M.H. A and M.H. B.

Depth of M.H. A = (100.000 − 99.000)	=	1.000
Depth of M.H. B	=	1.425
		2.425 m

$$\text{average depth of trench} = \frac{2.425}{2} = \underline{1.213 \text{ m}}$$

It is common practice in taking-off to use a schedule to calculate all relevant
invert levels and average depths of sewer trenches before setting down the
totals on dimension paper.

Measurement of irregular areas and use of compensation lines

To find the area of a composite shape such as a building site we may proceed by
dividing the area into triangles. The total area can then be found by calculating

the area of each individual triangle using ½ base × height.
For example the area of the triangle ABE in Fig. 6.8 would be:

$$\text{area ABE} = \tfrac{1}{2}\,(30 \times 10)$$

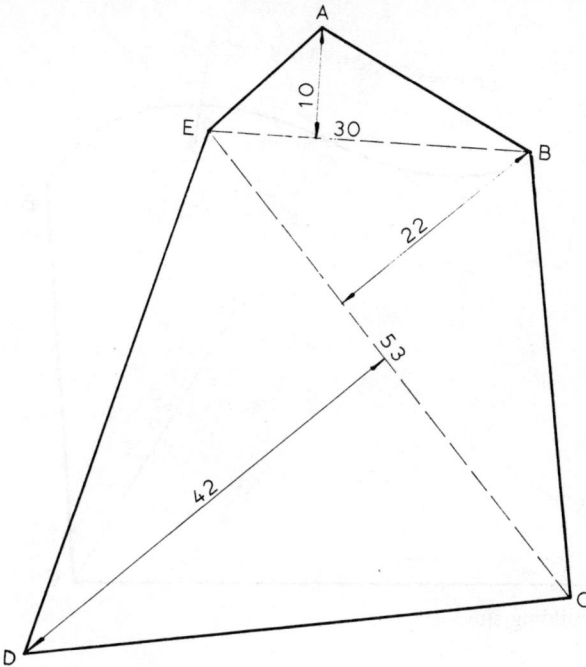

Fig. 6.8 Plan of building site

Example 6.3

½/	30.00		⌉ Area	(ABE)
	10.00	150.00		
½/	53.00			(BCE)
	22.00	583.00		
½/	53.00			(CDE)
	42.00	1113.00 ⌋		
		1846.00		

The total area of the site in Fig. 6.8 may be set down on dimension paper as in Example 6.3.

Figure 6.9 indicates a plan of a building site which is irregular and also has one boundary of a curved nature. The quantity surveyor can still make use of triangulation to calculate the area of the site by drawing a compensation line

72   through the curved boundary, in effect making it straight. Compensation lines, often known as give and take lines for obvious reasons, should be drawn as accurately as possible. The area of Fig. 6.9 would be set down on dimension paper as shown in Example 6.4.

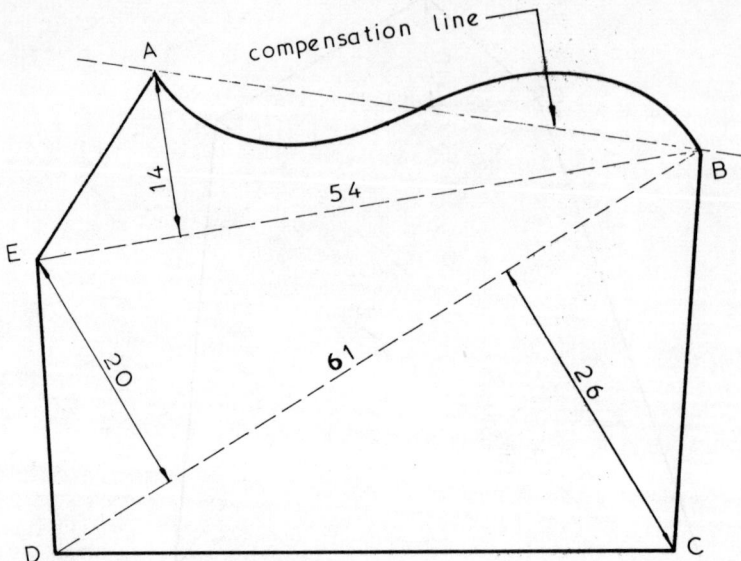

Fig. 6.9 Plan of building site

Example 6.4

½/	54.00		Area	(ABE)
	14.00	378.00		
½/	61.00			
	20.00	610.00		(BDE)
½/	61.00			
	26.00	793.00		(BCD)
		1781.00		

Trigonometry and further uses of Pythagoras

We saw earlier how we could use Pythagoras to calculate the sloping length of gradients. In the measurement of roofing work we can make further use of the theorem and also of simple trigonometry. There are three measurements which are commonly required when measuring roofing work. These are:

1. length of common rafters (used for length of rafters and area of tiling);
2. rise of roof;
3. length of hip rafters (on hip-ended roofs only).

Where there is a large-scale drawing available the lengths required can be
scaled from the drawing. Calculation is, however, fairly simple and gives
greater accuracy. Students will normally be familiar with the construction of
pitched roofs at this stage of their studies, but Fig. 6.10 is included to show the
relative positions of the common rafters and hip rafters in a hipped roof.

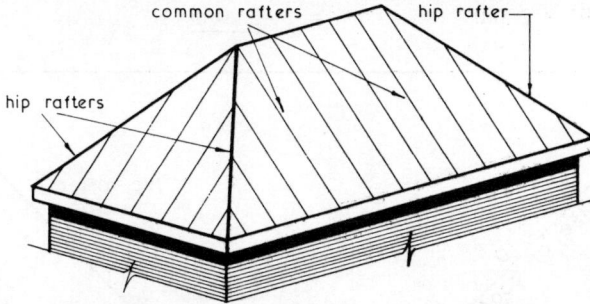

Fig. 6.10 Hip-ended roof

Example 6.5 indicates how trigonometry and Pythagoras may be used to
calculate the lengths previously described.

Example 6.5. Figure 6.11 shows the section through half of the roof. The
pitch (angle) of the roof is 40° and this is always given on the drawings.

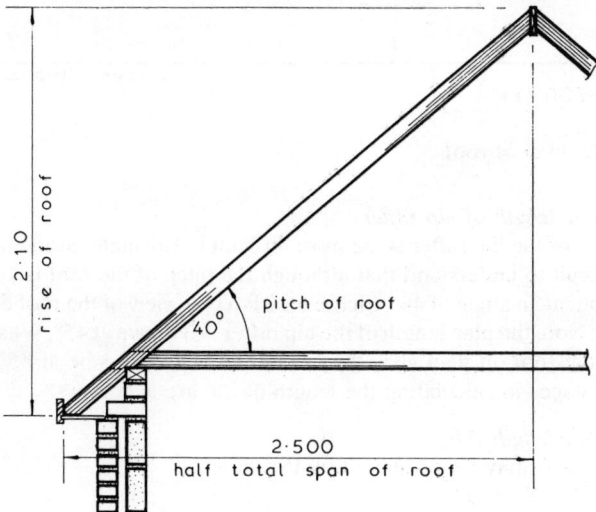

Fig. 6.11 Section through roof

To calculate length of rafter
Use secant of angle of pitch × half total span of roof. (Note total span is
measured from extreme overhang of eaves to centre of roof.)

in Figure 6.11 length of rafter =
Secant 40° × 2.500 =
1.305 × 2.500 = <u>3.26 m</u>

To calculate rise of roof
We need to know the rise of the roof in order that we may calculate the length
of the hip rafter.

Use tangent of angle of pitch of roof × half total span of roof.

in Fig. 6.11 rise of roof   =
Tangent 40° × 2.500   =
0.8391 × 2.500   =   **2.10 m**

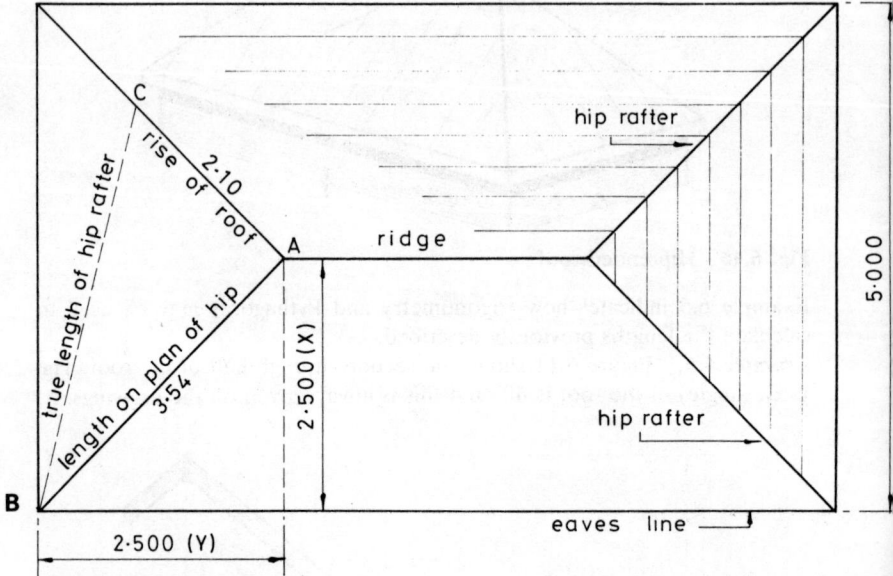

Fig. 6.12 Plan of roof

To calculate length of hip rafter
The length of the hip rafter is the most difficult to calculate. Students find it a
little difficult to understand that although the pitch of the roof is 40° the hip
rafter is not at an angle of 40°. Figure 6.12 is a plan view of the roof detailed in
Fig. 6.11. Note the plan length of the hip rafter AB drawn at 45°. Whatever the
pitch of the roof on plan view the hip rafters will always be at 45°.

The stages in calculating the length of hip are as follows:

1. *Calculate length AB*
   Length AB may be calculated by Pythagoras.
   $AB = \surd\,(X^2 + Y^2)$
   $AB = \surd\,(6.25 + 6.25)$
   $AB = \surd\,12.50$
   Length AB = 3.54

2. *Calculate rise of roof*
   This was previously calculated as 2.10 m. The rise may be indicated on Fig.
   6.12 as length AC. The dotted line BC therefore becomes the true length of
   the hip rafter. This may once again be calculated using Pythagoras.
   $BC = \surd\,(AB^2 + AC^2)$
   $BC = \surd\,(3.54^2 + 2.10^2)$

BC = √ (12.53 + 4.10)
BC = √ 16.63
    length of hip rafter BC = <u>4.08 m</u>

Measurement overall and adjustment of wants and voids

When measuring building work every attempt must be made to be as accurate as possible with the final quantities. To help in this aim it is the practice to measure quantities full in the first instance and make deductions for such items as doorways and window openings later. Such deductions are called adjustments. To illustrate this procedure let us measure the simple area of the room detailed in Fig. 6.13. Many students would probably attempt to measure

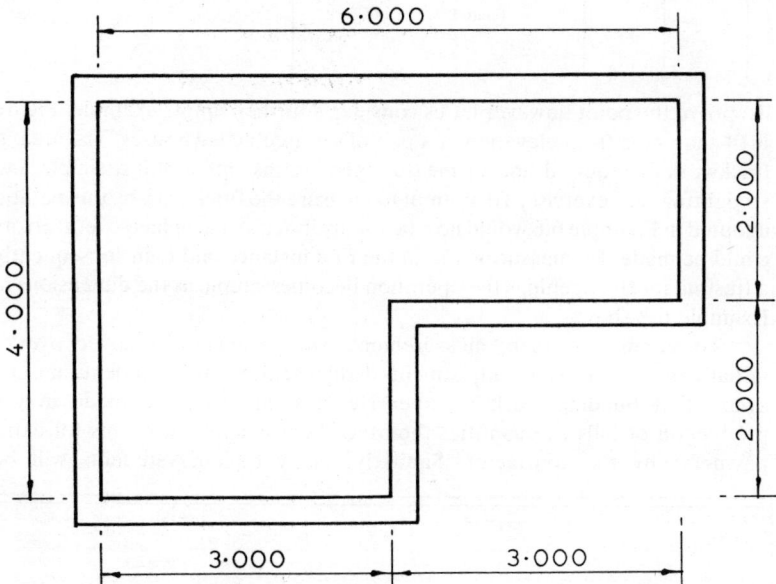

Fig. 6.13 Plan of room

this particular area as shown in Example 6.6. Although the correct answer for the area of Fig. 6.13 has been arrived at in Example 6.6, the better way to

Example 6.6				
	3.00		Area Fig. 6.13	Here Fig. 6.13 has been measured as two rectangles to arrive at the total area.
	4.00	12.00		
	3.00			
	2.00	6.00		
		18.00		

measure would be as shown in Example 6.7. Example 6.7 may seem an unnecessarily complicated way of measuring the area of such a simple shape.

Example 6.7

6.00		⌐ Area Fig. 6.13
4.00	24.00	⌐
3.00		⌐ *Deduct*
2.00	6.00	⌐
		ditto
		(want)

Here Fig. 6.13 is measured as a full rectangle 6.00 X 4.00 and is then adjusted by deducting the corner (known as a want) size 3.00 m X 2.00 m.

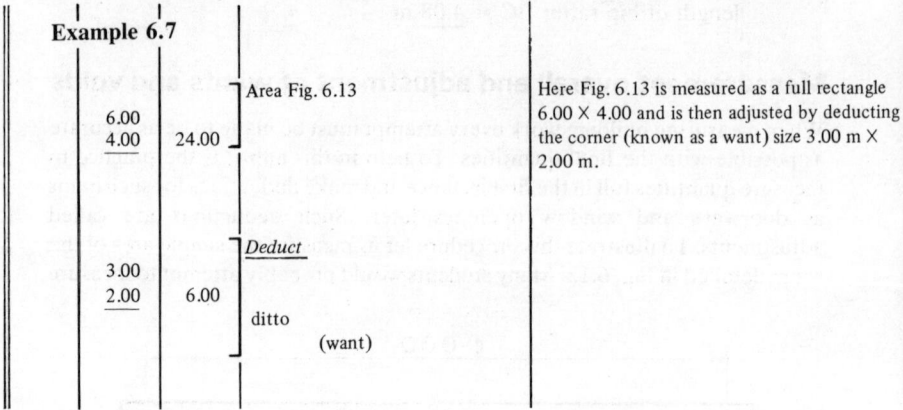

To prove the point however let us consider a further simple example. Figure 6.14 shows the front elevation of a pair of semi-detached houses. The area of brickwork is required and in measurement terms this is still a simple and straightforward exercise. To attempt to measure the brickwork by the method adopted in Example 6.6 would now be totally illogical and in fact several errors could be made. By measuring full in the first instance and then subsequently adjusting for the openings the operation becomes simple as the dimensions in Example 6.8 show.

To summarize, then, measurements should always be made overall initially and any necessary adjustments deducted afterwards. Students must be aware that building work is expensive. Should errors be made in the production of bills of quantities then the client is liable to claims for extra payments by the contractor. Similarly, many building students will be

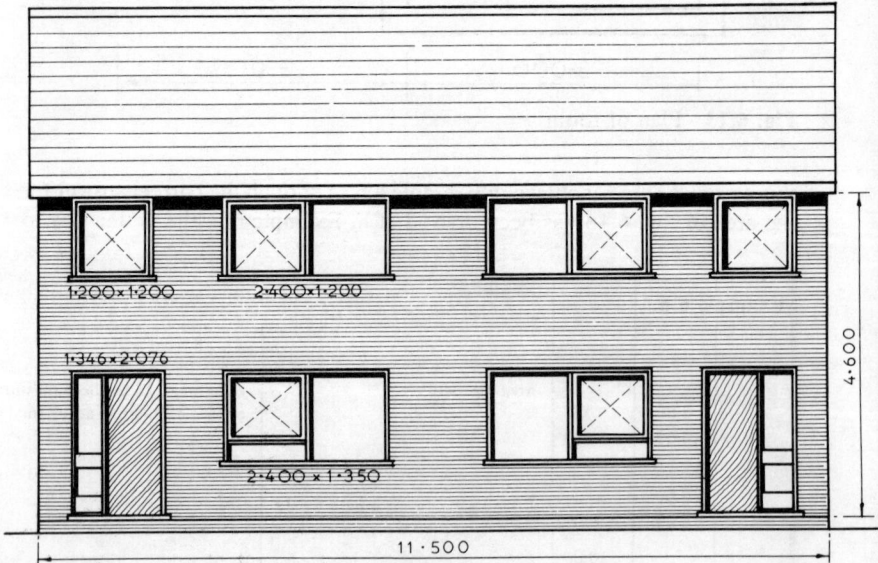

Fig. 6.14 Front elevation of semi-detached houses

Example 6.8

	11.50		Area of brickwork
	4.60	52.90	
2/	1.35		*Deduct*
	2.08	5.62	ditto
2/	2.40		
	1.35	6.48	
2/	2.40		
	1.20	5.76	
			(window and
2/	1.20		door openings)
	1.20	2.88	
		20.74	

responsible for measuring for ordering purposes. Others will be measuring quantities for the pricing of building works. In either case errors could cost your company dearly.

By measuring overall and then deducting for adjustments the likelihood of error, if not eliminated, is at least reduced.

Wants and voids

These two terms are used in measurement work in the following manner.

Wants

A want is an opening or a break along the boundary of an area, room, wall, etc. being measured. Wants are always deducted whatever their size. Figure 6.15 illustrates typical wants.

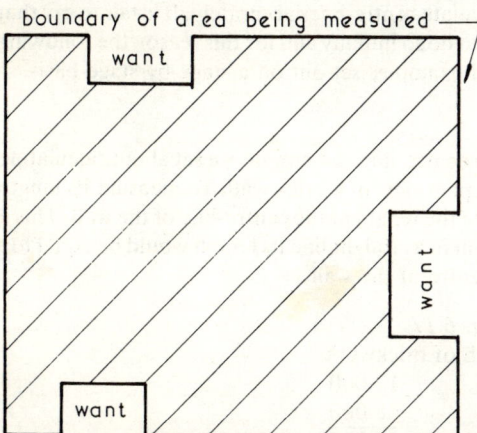

Fig. 6.15 Typical wants

Voids

A void differs from a want in that it is wholly within and detached from the boundaries of the area being measured. Voids are not always deducted when measuring for bills of quantities. The *Standard Method of Measurement* lays down the minimum size a void must be before any deduction is made to the area being measured. For example:

1. no deduction in brickwork shall be made for a void not exceeding 0.10 m²;
2. no deduction on formwork shall be made for voids not exceeding 5.00 m²;
3. no deduction of timber boarding and flooring shall be made for voids not exceeding 0.50 m².

Figure 6.16 illustrates typical voids

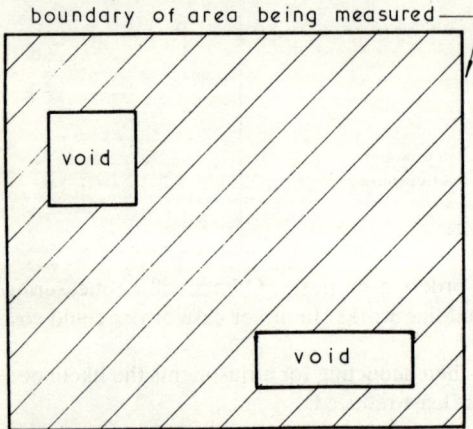

Fig. 6.16   Typical voids

Centre-line calculation

Centre-lines are used extensively in the measurement of the structure of a building. Examples are excavation of foundation trenches, backfilling to trenches, temporary support to foundation trenches, concrete in foundations, brickwork and blockwork. Use will be made of centre-line calculations in the worked measurement exercises later in the book. It is important, therefore, that the student be able to calculate centre-lines accurately. It is fair to say that many students find it difficult to do so initially and for this reason the following pages contain several worked examples set out on a stage-by-stage basis.

What are centre-lines?

What exactly do we mean by a centre-line and how do we set about calculating them? Figure 6.17 details the plan view of a brick wall. To measure its length accurately we need to calculate the length of the centre-line of the wall. This is because if the wall were laid out in a straight line its length would be equal to a line measured through the centre of the wall.

To calculate centre-line of Fig. 6.17
The external perimeter length of brickwork

from drawing	= 2/6.000	12.000
	= 2/4.000	8.000
		20.000

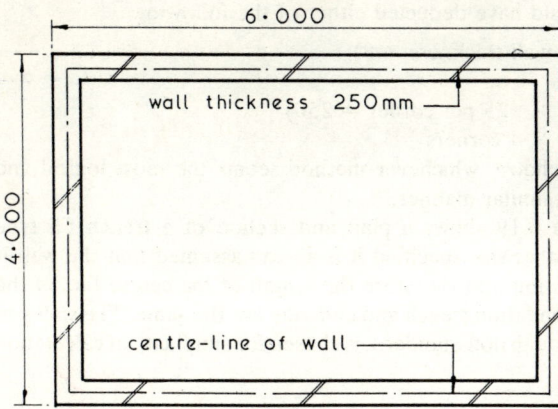

Fig. 6.17 Centre-line calculation

Consider plan of one corner Fig. 6.18.

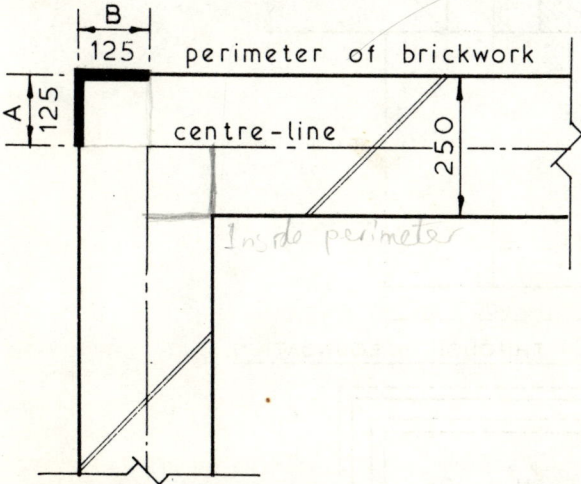

Fig. 6.18 Plan of one corner

Clearly at this one corner the perimeter of the brickwork is longer than the centre line of the wall by A + B.

There are four corners in this wall so if we were to deduct length A + B × the four corners from the perimeter of the brick wall we would have calculated the length of the centre-line.

The wall thickness is given as 250 mm wide. Both dimensions A and B must equal half the total wall thickness.

centre-line of wall	=	
Perimeter of brickwork as before	=	20.000
Deduct		
4/2/125 (2 × 125 per corner)	=	1.000
and four corners		
centre-line of wall	=	<u>19.000 m</u>

Alternatively, we could have deducted either of the following:

8/125 $=$ 1.000 ($\frac{1}{2}$ wall thickness $\times$ 8)

or

4/250 $=$ 1.000 (2 $\times$ 125 per corner $=$ 250)
$\times$ 4 corners

The student should choose whichever method seems the most logical and always calculate in a similar manner.

Example 6.9 Figure 6.19 shows a plan and section of a trench fill type foundation. Unless otherwise specified it is always assumed that the wall is central of the foundation and therefore the length of the centre-line of the brick wall and the foundation trench and concrete are the same. The wall size of 250 mm wide, although non-standard, is chosen for simplicity of calculation in this first example.

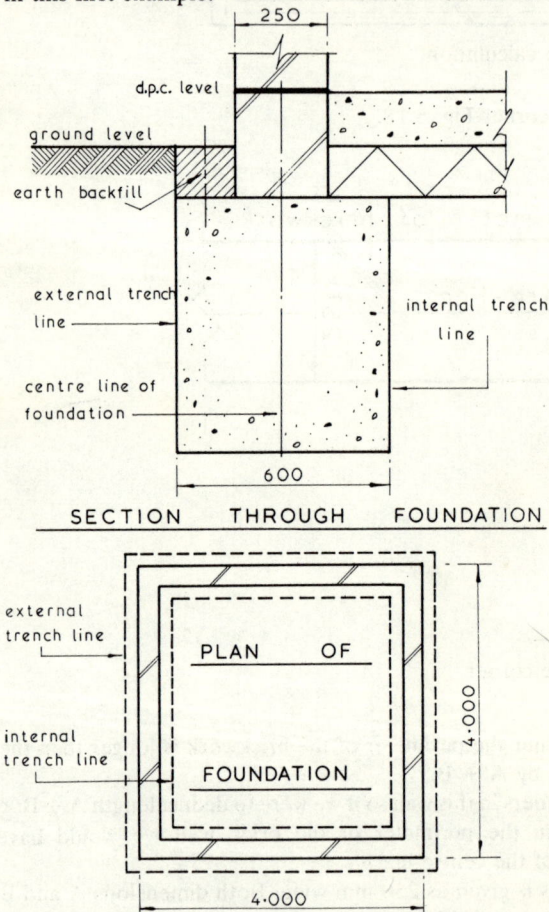

Fig. 6.19 Plan and section of trench fill foundation

The following lengths may be required for measurement purposes:

1. centre-line of the brick wall and foundation;
2. length of external trench line;
3. length of internal trench line;

4. centre-line of earth backfilling.

1. To calculate centre-line of brick wall and foundation
Consider Fig. 6.20

Perimeter of brickwork from plan =	4/4.000	= 16.000
Deduct 4/2/125		1.000
Centre-line length	=	15.000

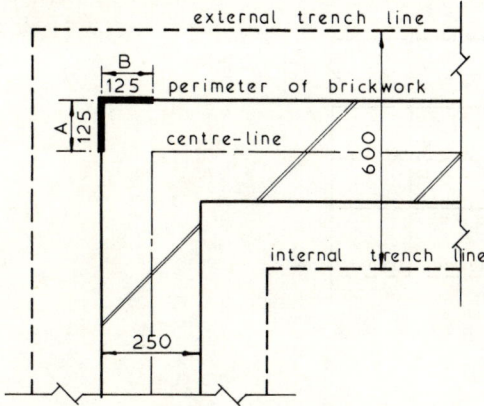

Fig. 6.20 Calculation in centre-line of brick wall and foundation

2. To calculate length of external trench line

Consider Fig. 6.21. Clearly the length of external trench line is longer than the centre-line by A + B at each corner.

external trench line =

Centre-line of foundation as before	=	15.000
Add 4/2/300 (2 × 300 per corner)	=	2.400
× 4 corners		
length of external trench line	=	17.400

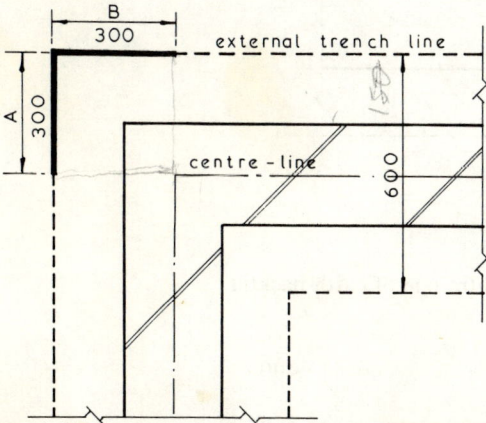

Fig. 6.21 Calculation of external trench length

3. *Length of internal trench line*
Consider Fig. 6.22
Clearly this is the opposite situation of the external trench line.

Centre-line as before	=	15.000
Deduct 4/2/300	=	2.400
length of internal trench line		12.600

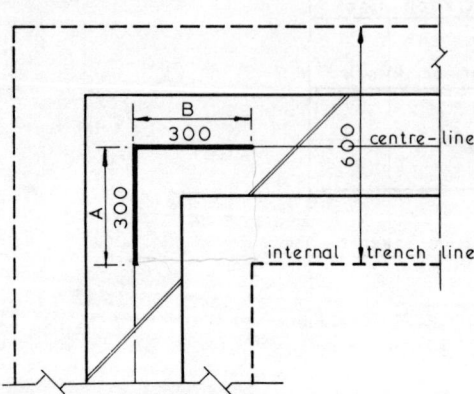

Fig. 6.22 Calculation of internal trench length

4. *Centre-line of earth backfill*
Consider Fig. 6.23
The centre-line of the earth backfill is central between the external trench line and the perimeter of the brick wall. Its length, therefore, must be the average of those two.

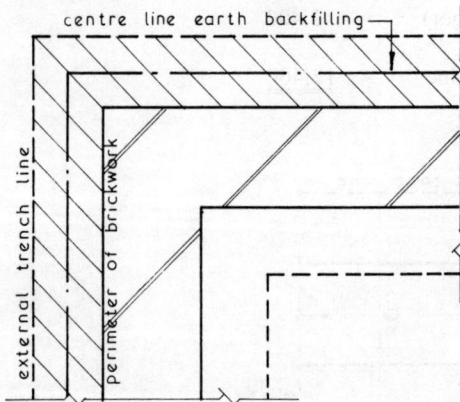

Fig. 6.23 Calculation of centre-line of earth backfill

Length of external trench line	=	17.400
Length of perimeter of brickwork	=	16.000
	=	33.400

centre-line of earth
backfill = $\dfrac{33.400}{2}$ = 16.700

Cavity walls

Probably the most difficult centre-lines to calculate are those involving cavity walls where the individual skins are a different thickness. Figure 6.24 indicates a section and plan of such a foundation. For simplicity a trench fill style foundation is used.

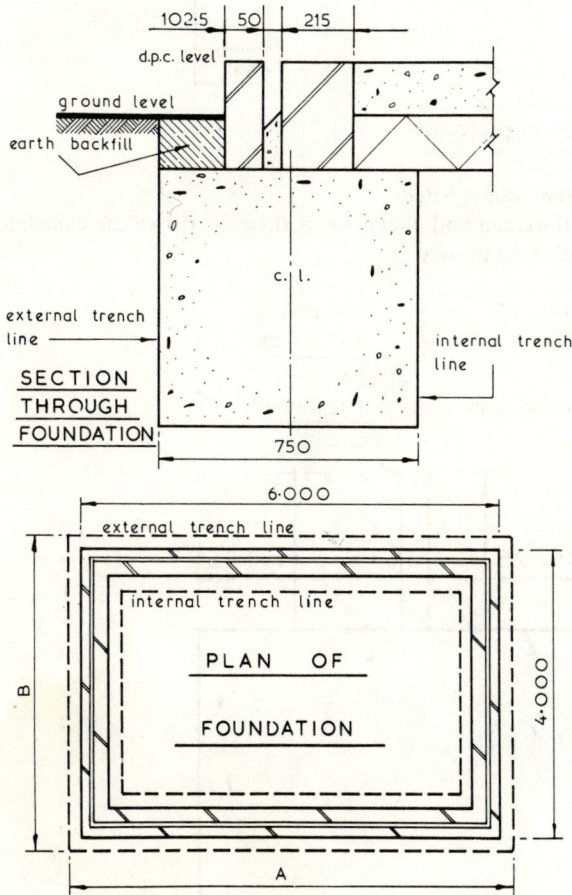

Fig. 6.24 Plan and section of trench fill foundation

For measurement purposes we need to calculate the following:

1. length of external trench line;
2. centre-line of foundation trench;
3. length of internal trench line;
4. centre-line of external brick wall;
5. centre-line of cavity;
6. centre-line of internal brick wall;
7. centre-line of earth backfill.

Example 6.10 In this example we will use different methods to those used in Example 6.9 to indicate to the student that alternative methods may be adopted. Again we are to assume that the mass of the whole wall is sitting central on the foundation concrete.

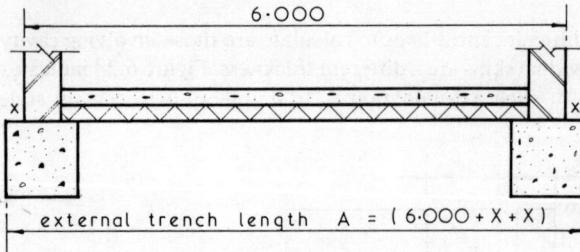

Fig. 6.25 Calculation of projection

1. To calculate external trench length
Consider Fig. 6.25. If we can find dimension X then clearly we can calculate external trench length for one side.
To find X

Width of foundation	=	750 mm
Width of wall	=	367.5 mm
	2)	382.5 mm
projection of foundation	=	191.25 mm

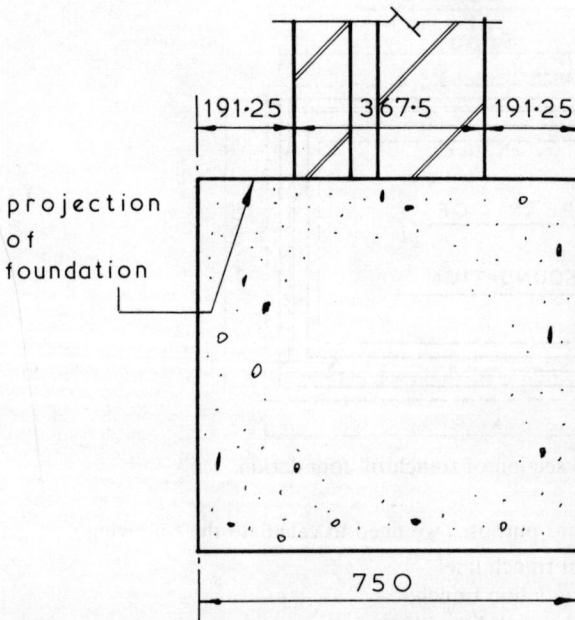

Fig. 6.26 Calculation of external trench length

Length of external trench length A = (6.000 + 2/191.25) =   6.383
Length of external trench length B = (4.000 + 2/191.25)   =   4.383

Total length of external trench	= 2/6.383	=	12.766
	2/4.383	=	8.766
			21.532

2. To calculate centre-line of foundation trench
Consider Fig. 6.27.

External trench perimeter as before	=	21.532
centre-line of foundation deduct corners 4/2/375	=	3.000
		18.532

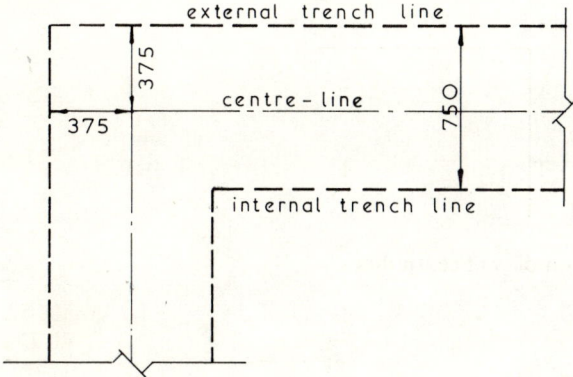

Fig. 6.27 Calculation of centre-line of trench

3. To calculate internal trench line
Consider Fig. 6.28

External trench length as before	=	21.532
internal trench length deduct corners 4/2/750	=	6.000
		15.532

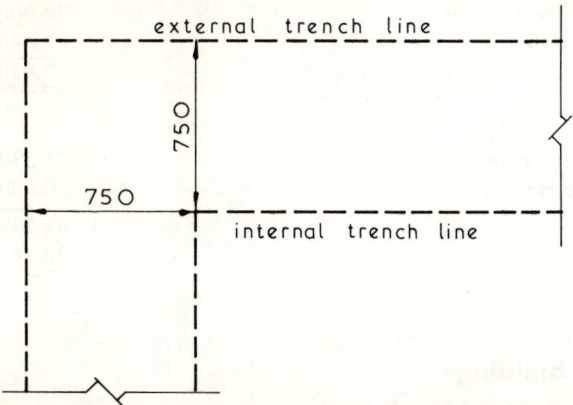

Fig. 6.28 Calculation of internal trench length

86 **Wall calculations**
Consider Fig. 6.29.

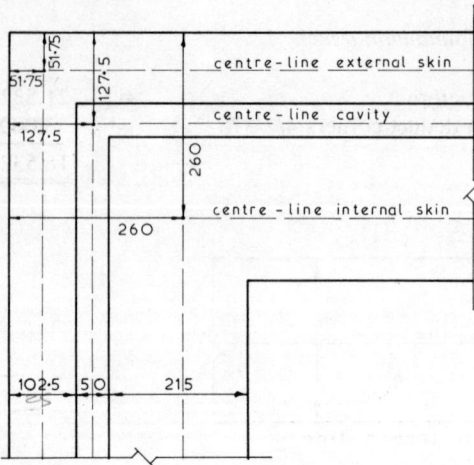

Fig. 6.29 Calculation of wall centre-lines

4. Centre-line external skin

Perimeter of brickwork 2/6.000 + 2/4.000	=	20.000
Deduct corners 4/2/51.75	=	0.410
		19.590

5. Centre-line of cavity

Perimeter of brickwork as before	=	20.000
Deduct corners 4/2/127.5	=	1.020
		18.980

6. Centre-line of internal skin

Perimeter of brickwork as before	=	20.000
Deduct corners 4/2/260	=	2.080
		17.920

7. Centre-line of earth backfill

External trench length as before	=	21.532
Add perimeter of brickwork	=	20.000
Average =	2)	41.532
		20.766

Plan shapes of buildings

So far we have considered only those plan shapes that are square or
rectangular. Figures 6.30 and 6.31 indicate two plan shapes that at first sight

do not appear to be rectangular. To calculate the centre-lines of either of these   87
buildings we would need to know the following:

1. the length of the external perimeter of the wall;
2. the number of corners that need to be adjusted to calculate centre-line
   lengths.

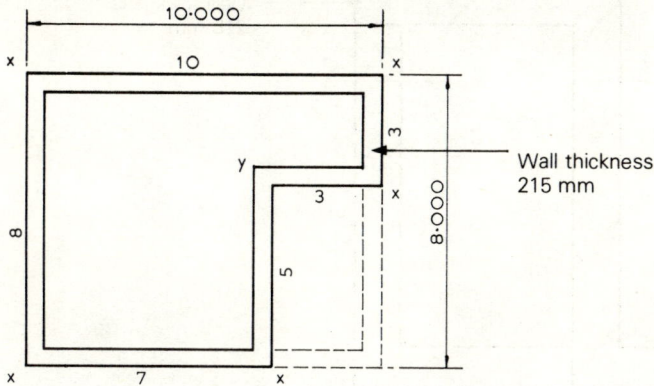

Fig. 6.30   Plan of building

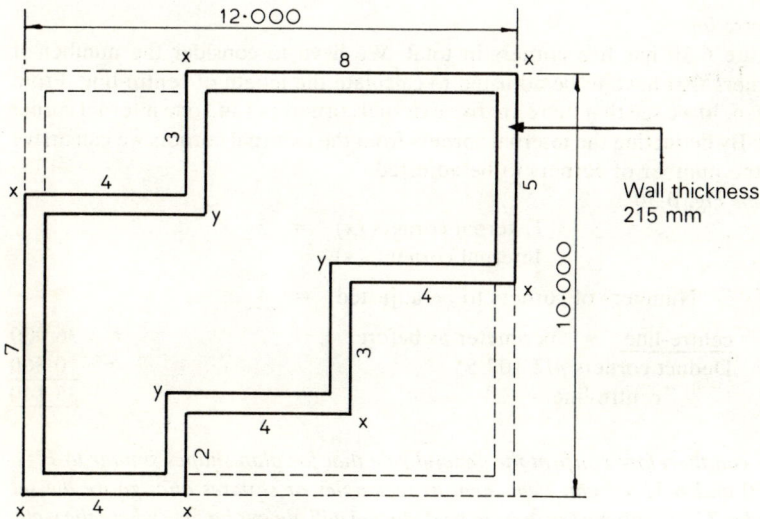

Fig. 6.31   Plan of building

Consider Fig. 6.30:   If we measure around the building in a clockwise
direction we can calculate the external perimeter length of the wall:

$$10 + 3 + 3 + 5 + 7 + 8 = 36.000 \text{ m}$$

Alternatively, if we calculate the perimeter of wall by using the overall
dimensions we find that we arrive at the same answer:

2/10.000   =   20.000
2/ 8.000   =   16.000

Perimeter   =   36.000 m

Clearly, the length of the set back must be the same as the rectangle formed by the dotted line, see Fig. 6.32.

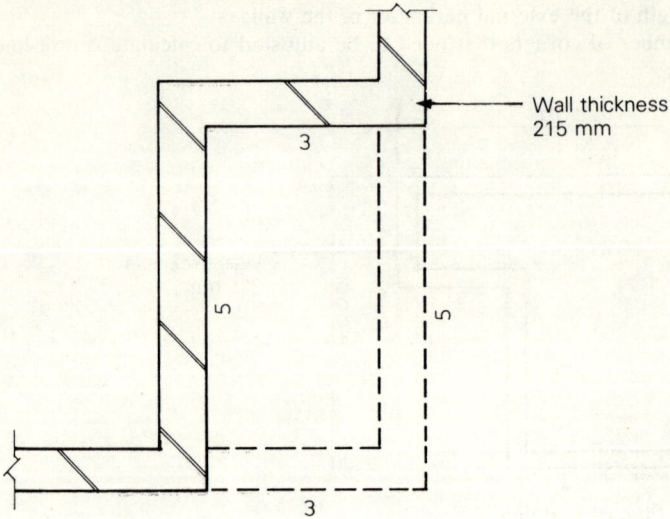

Fig. 6.32 Detail of corner for Fig. 6.30

Centre-line
Figure 6.30 has five corners in total. We have to consider the number of corners that have to be adjusted to calculate the length of centre-line. From Fig. 6.30 we see that there are five external corners (x) and one internal corner (y). By deducting the internal corners from the external corners we can arrive at the number of corners to be adjusted.
From Fig. 6.30:

$$
\begin{array}{lcl}
\text{External corners (x)} & = & 5 \\
\text{Internal corners (y)} & = & \underline{1} \\
\text{Numbers of corners to be adjusted} & = & \underline{4}
\end{array}
$$

centre-line = perimeter as before	=	36.000
Deduct corners 4/2/107.5	=	0.860
centre-line =		35.140

We can therefore confirm the general rule that for plan shapes similar to Figs. 6.30 and 6.31 we can treat them as rectangles or squares and ignore the set backs. The number of corners to be deducted will always be four where the walls join up to enclose a building.
To prove these observations let us consider Fig. 6.31.

Perimeter
Again measuring around building in a clockwise direction:
$8 + 5 + 4 + 3 + 4 + 2 + 4 + 7 + 4 + 3 = 44.000$ m
Alternatively, using overall dimensions,

2/12.000	=	24.000
2/10.000	=	20.000
the same as before		44.000 m

Centre-line

External corners (x)	=	7
Deduct internal corners (y)	=	3
Number of corners to be adjusted	=	4
Perimeter as before	=	44.000
Deduct corners 4/2/107.5	=	0.860
centre-line	=	43.140 m

Projections

Buildings often have projections along the length of a wall. Figure 6.33 is a typical example. To arrive at the perimeter of this building we can mentally move the shaded portion of wall length A into the position shown dotted. This leaves the two lengths shaded black which must be added to arrive at the total perimeter.

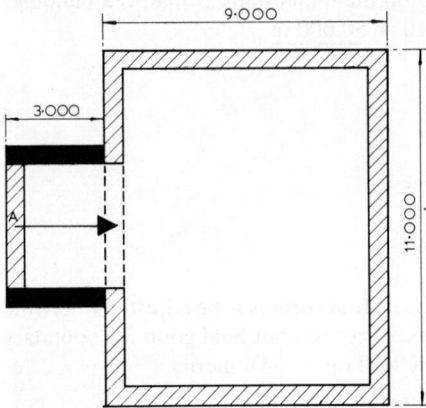

Fig. 6.33   Plan of building

For example:

Perimeter	=	2/ 9.000	=	18.000
		2/11.000	=	22.000
Add projection		2/ 3.000	=	6.000
External perimeter length of building			=	46.000

Centre-lines should then be calculated in the normal way deducting four corners.

A further example of this type of projection is given in Fig. 6.34. Again we can make the shape rectangular by mentally slotting the shaded areas into the dotted positions by following the direction of the arrows. We are then left with the two black areas of wall which must be added to the overall rectangular size.

external wall perimeter	=	2/13.000	=	26.000
		2/10.000	=	20.000
Add black shaded areas		2/ 2.000	=	4.000
External perimeter length of building			=	50.000

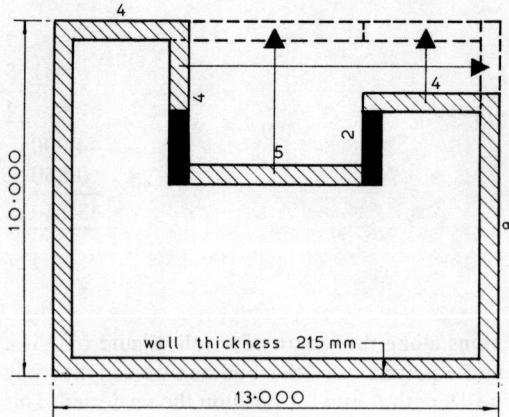

Fig. 6.34 Plan of building

Alternatively, we could simply add up the dimensions around the building:
4 + 4 + 5 + 2 + 4 + 8 + 13 + 10 = 50.000 m

Centre-line

External corners	= 6
Internal corners	= 2
number of corners to be adjusted	= 4

Boundary wall

The general rule that there will always be four corners to be adjusted where the walls join to enclose a building does, of course, not hold good for boundary walls. We must treat each boundary wall on its own merits.

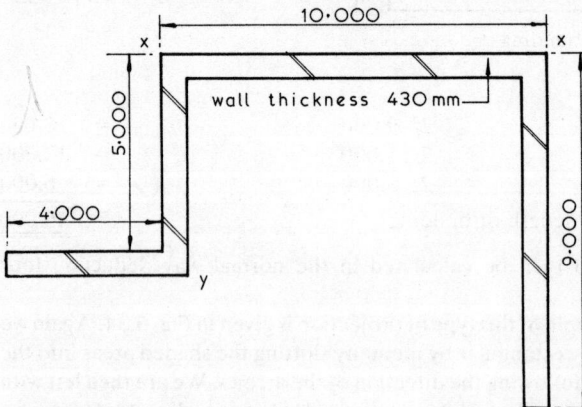

Fig. 6.35 Boundary wall

Consider Fig. 6.35:
Length of wall from figured dimensions = 4 + 5 + 10 + 9 = 28.00 m
∴ centre-line

Number of external corners (x)	=	2
Number of internal corners (y)	=	1
∴ to calculate centre-line length one corner must be adjusted		1
∴ wall length as above	=	28.000
Deduct corner 2/215	=	0.430
(wall thickness = 430 mm)	=	27.570
∴ centre-line = 27.570 m		

Conclusion

Students for one reason or another find difficulty generally with centre-line calculation. The author strongly advises students when calculating centre-lines to draw a plan view sketch of the corner on a rough piece of paper, similar to the drawings used in this book. This helps the student to visualize the problem.

Interpolation of levels

It is necessary for students to make use of interpolation of levels in measurement as well as in surveying.

Frequently this will involve the calculation of the volume of excavation over a building site. It is common for building sites to be of a sloping nature and excavation and filling is undertaken to make the site flat. The quantity surveyor has to calculate the amount of excavation and of filling material required. To enable this to be done a grid of regularly spaced spot ground levels is prepared. The level to which the site is to be excavated down to is known as the reduced level. Figure 6.36 shows a section through a building site, the reduced level required being 100.00.

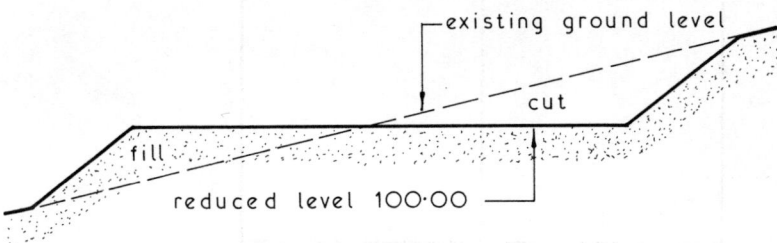

Fig. 6.36 Section through building site

Let us assume that a grid of levels has been prepared for a site as shown in Fig. 6.37. The reduced level required is 100.00. Soil above the 100.00 line would need to be excavated, and below the 100.00 line fill material would be required.

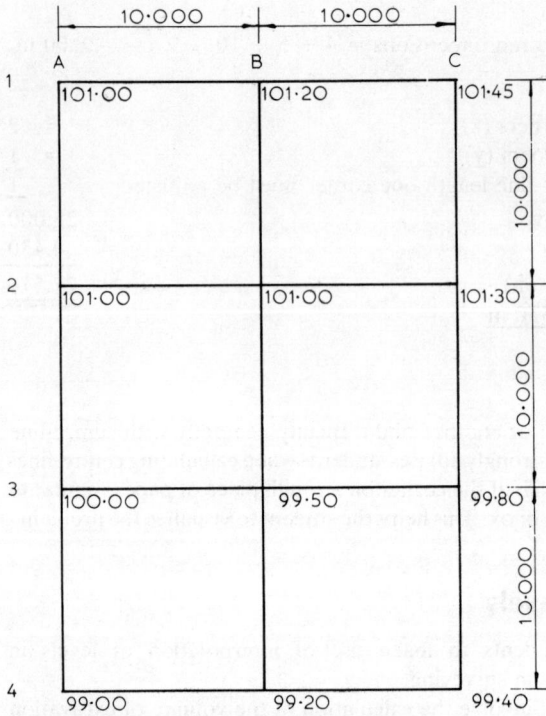

Fig. 6.37 Grid of levels

To measure the quantity of excavation we need to be able to plot the contour line 100.00. Clearly this passes through the middle squares and its position can be calculated from Fig. 6.38.

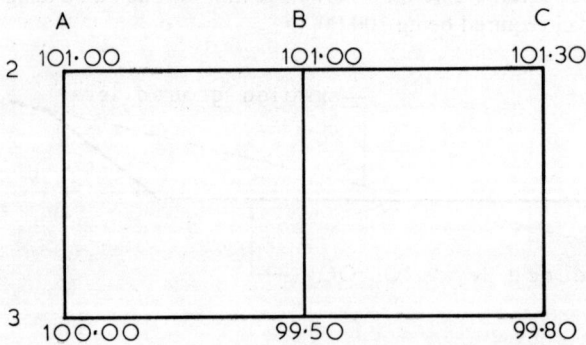

Fig. 6.38 Part of grid

Position of contour line 100.00 on grid line B2–B3

$$\text{Length of grid line} \left(\frac{\text{reduced level} - \text{B3}}{\text{B2} - \text{B3}} \right) =$$

$$= 10 \left(\frac{100.00 - 99.50}{101.00 - 99.50} \right) = 10 \times \frac{0.50}{1.50} = 3.33 \text{ m}$$

$$10 \left(\frac{100.00 - 99.80}{101.30 - 99.80} \right) = 10 \times \frac{0.20}{1.50} = 1.33 \text{ m}$$

We can now plot the reduced level line of 100.00 as in Fig. 6.39.

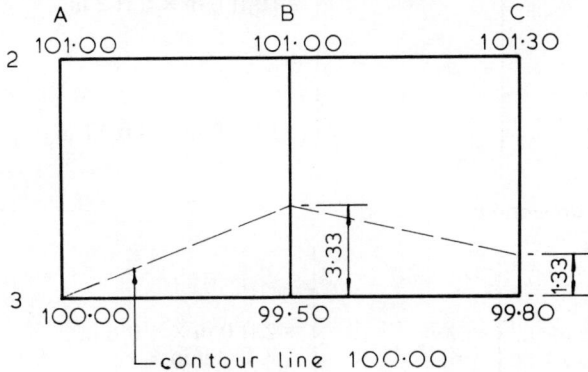

Fig. 6.39 Contour line 100.00

It is necessary at this level for the student to be able to interpolate levels only. Worked measurement examples will be undertaken at Level III. The student is also directed to the measurement example of excavation to reduce levels in Example 8.1 (p. 104) of this book.

Imperial measurement

Prior to the introduction in 1968 of metrication in the construction industry measurement work in general was undertaken in imperial units. Although the construction industry now works mostly in metric units there are times when imperial units are still used.

The following information is included to allow the student to compare the use of imperial units with metric methods described earlier in the book. In doing so the student will become aware of the problems associated with imperial units of meausrement and may wonder why the country as a whole did not change to the metric system earlier than it did.

Taking-off in imperial units

The dimensions are recorded on dimension paper in feet and inches. Fractions of an inch such as a ¼ or ½ an inch are used in waste calculations to calculate total dimensions. They are disregarded in the dimension column itself.

Examples of linear dimensions

5·6		means 5 ft 6 in
22·10		means 22 ft 10 in
30·9		means 30 ft 9 in

$$
\begin{array}{l}
10 \cdot 6 \\
\underline{2 \cdot 2}
\end{array}
$$

= 10 ft 6 in × 2 ft 2 in

$$
\begin{array}{l}
12 \cdot 9 \\
\underline{4 \cdot 11}
\end{array}
$$

= 12 ft 9 in × 4 ft 11 in

Examples of cubic dimensions

$$
\begin{array}{l}
2 \cdot 0 \\
4 \cdot 6 \\
2 \cdot 9
\end{array}
$$

= 2 ft 0 in × 4 ft 6 in
  × 2 ft 9 in

$$
\begin{array}{l}
2 \cdot 11 \\
3 \cdot 1 \\
2 \cdot 3
\end{array}
$$

= 2 ft 11 in × 3 ft 1 in
  × 2 ft 3 in

Squaring of imperial dimensions

In setting down dimensions in feet and inches when we come to squaring up we have to realize that the inches are twelfth parts and not tenths as they would be with traditional decimal calculations.

For example in writing down a dimension as 12.9 we mean 12%12. This is not the same as 12.9 in traditional decimal form which expressed as a fraction would be 12%10.

Before electronic calculators came into common use squaring of dimensions had to be carried out manually.

A system known as duo-decimals was commonly used for this purpose.

Duo-decimal calculation

Assume that we are to square up the following dimension

$$
\begin{array}{l}
4 \cdot 6 \\
\underline{2 \cdot 3}
\end{array}
$$

= 4 ft 6 in × 2 ft 3 in

Stage 1. Multiply out (feet × feet = feet)

$$
\begin{array}{l}
4.6 \\
\underline{2.3} \quad = \quad 4 \times 2
\end{array}
$$

= 8.0

Stage 2. Multiply (feet × inches = twelfths)

4.6

↗

2.3 = 2 × 6 = 12 ÷ 12   = 1.0

4.6

↘

2.3 = 4 × 3 = 12 ÷ 12   = 1.0

Stage 3. Multiply (inches × inches = parts)

4.6

2.3 = 6 × 3 = 18 ÷ 12   = 0.1.5

total =   8.0

1.0

1.0

1.5

10.1.5

This means 10ft 1in and 5/12 parts. The parts are rounded to the nearest inch. Had the parts been 6/12 or over then the answer would be 10ft 2in. Let us consider one further example.

4.9

3.8

feet × feet   4 × 3 =   12.0

feet × inches 4 × 8 = 32 ÷ 12   =   2.8

feet × inches 3 × 9 = 27 ÷ 12   =   2.3

inches × inches 8 × 9 = 72 ÷ 12   =   0.6

17.5

With practice duo-decimals are not as difficult as might appear from these examples although cubic calculations take a little longer. With the advent of electronic calculators the manual squaring process has largely become redundant. Let us now consider squaring imperial dimensions using a calculator.

Unit inches	Fraction of a foot	Decimal equivalent
1	$\frac{1}{12}$	0.083
2	$\frac{2}{12}$	0.167
3	$\frac{3}{12}$	0.250
4	$\frac{4}{12}$	0.330
5	$\frac{5}{12}$	0.417
6	$\frac{6}{12}$	0.500
7	$\frac{7}{12}$	0.583
8	$\frac{8}{12}$	0.667
9	$\frac{9}{12}$	0.750
10	$\frac{10}{12}$	0.833
11	$\frac{11}{12}$	0.917

Table 6.1

Squaring imperial dimensions by calculator

The first essential is to prepare the decimal equivalents of the inch units of measure 1 to 11 inches inclusive. These are shown in Table 6.1. Such a table is quickly prepared and should be kept handy where much squaring of imperial dimensions is contemplated.

As examples let us use the dimension previously squared by duo-decimals:

			From Table 6.1
4·6			0.6 = 0.5
2·3			0.3 = 0.25
			By calculator
			4.50 × 2.25 = 10.125

We must refer again to Table 6.1, 0.125 is nearest to 0.083 and is therefore 1in. The squared answer is therefore 10ft   1in

			From Table 6.1
4·9			0.9 = 0.750
3·8			0.8 = 0.667

4.75 × 3.667 = 17.418
Referring to Table 6.1, 0.418 is nearest to 0.417 = 5in
The squared answer is therefore 17ft 5in

Reduction

Although the dimensions are measured and squared in feet and inches they are reduced to linear, square or cubic metres for inclusion into bills of quantities. These are also convenient sizes for ordering purposes. Before squared dimensions are reduced the totals of each measured item should be rounded up or down to the nearest yard.

Linear yards	=	linear feet ÷ 3
Square yards	=	square feet ÷ 9
Cubic yards	=	cubic feet ÷ 27

Chapter Seven

The Standard Method of Measurement of Building Works

Reference has been made several times in this book to a publication known as the *Standard Method of Measurement of Building Works*. This is basically a reference book published under the sponsorship of the Royal Institute of Chartered Surveyors and National Federation of Building Trades Employers Committee. Its purpose is to lay down a common and uniform method of measuring building work. It applies equally to proposed and executed works. The edition currently used is Standard Method of Measurement 6th edition commonly known as SMM 6.

History of SMM

Prior to 1922 there were no specific rules for the way in which building work was measured or how disputes should be settled. Quantity surveyors often used different methods and contractors misinterpreted the meaning of descriptions in bills of quantities leading to inaccuracies in tendering.

The desirability of having a set of standard rules for the measurement of building work became more and more apparent. Eventually the surveyor's institutions of the time set up a joint working committee whose task was to draw up a standard set of measurement rules. In 1918 representatives of the building industry were added to the committee. Then in 1922 the first *Standard Method of Measurement* was published.

By this time the committee had put in a great deal of effort and time in publishing the document. It was felt that rather than disband the committee it should in fact continue in existence. Its expertise could be used in an arbitrational capacity to advise on measurement problems. The first *Standard Method of Measurement* was revised in 1927.

By 1931 the *Standard Method of Measurement* was so widely recognized as the authority on measurement methods that it was incorporated into the form of buildings contracts issued under the sanction of the Royal Institute of British Architects.

Since the first *Standard Method of Measurement* was published in 1922 there have been six revisions. Probably the most dramatic change in all these years has been the change-over from imperial to metric units of measurement. This was carried out when the fifth edition was in use and a further edition known as fifth edition (metric) was published. This was a straight conversion from imperial to metric units and no attempt was made to change principles of measurement.

The current edition is then the first method to be revised with the metric system fully in use in the building industry. Comparing the methods over the years, probably the most noticeable changes would be the greater number of enumerated items today. Over those years the building industry has changed from a trade basis where everything is manufactured on site to one where major standard components are made in factories and simply erected on site. The standard method of measurement has changed in recognition of the modern building industry.

Other methods of measurement

Where reference is made in this book to the *Standard Method of Measurement* this means the current edition sponsored by the Royal Institute of Chartered Surveyors (RICS) and National Federation of Building Trades Employers (NFBTE).

There are other methods of measurement and the two common ones are listed below:

1. *Code for the Measurement of Building Works in Small Dwellings*
   This is also sponsored by the RICS and NFBTE and is a simplified version of the SMM for use in the measurement of small dwellings.
2. *Standard Method of Measurement of Civil Engineering Works*
   This document published by the Institute of Civil Engineers is for the measurement of civil engineering work as opposed to building work.

Using the standard method of measurement

Function

To describe the function of the *Standard Method of Measurement* one really can do no better than quote from the introduction of the document itself.

> This standard method of measurement provides a uniform basis for measuring building works and embodies the essentials of good practice but more detailed information than is required by this document shall be given where necessary in order to define the precise nature and extent of the required work. This Standard Method of Measurement shall apply equally to both proposed and executed works.

Contractual significance

That the SMM is of considerable significance is clearly seen when reading through the conditions of contract of some of the standard forms associated with construction work.

For example in the *JCT Guide to the Standard Form of Building Contract*, 1980 edition (private and local authority editions with quantities) the following wording is found

> The Contract Bills, unless otherwise specifically stated therein in respect of any specified item or items, are to have been prepared in accordance

Edition published by the Royal Institute of Chartered Surveyors and the National Federation of Building Trades Employers; if in the Contract Bills there is any departure from the method of preparation referred to in the above clause or any error in description or in quantity or omission of items, then such departure or error shall not vitiate this contract but the departure or error shall be corrected and such correction shall be treated as if it were a Variation required by an instruction of the Architect/Supervising Officer.

These are important clauses because should the quantity surveyor omit to measure work that is required to be measured in accordance with the SMM then the contractor would be able to claim for extra payment for that work.

Format of the SMM
The SMM is divided into sections as follows:

section A   –   general rules
section B   –   preliminaries
section C to X   –   work sections

General rules
This section lays down general rules which are to be observed in the measurement and preparation of bills of quantities.

Preliminaries
In Chapter 2 when discussing bills of quantities we saw that the first part consists of the preliminaries section. The SMM describes the information that should be included within this section of a bill of quantities.

Work sections
The work sections C to X inclusive of the SMM divide building work into twenty trade classifications. These were listed fully in Chapter 2, but can usefully be repeated here:

(C) demolition;
(D) excavation and earthwork;
(E) piling and diaphragm walling;
(F) concrete work;
(G) brickwork and blockwork;
(H) underpinning;
(J) rubble walling;
(K) masonry;
(L) asphalt work;
(M) roofing;
(N) woodwork;
(P) structural steelwork;
(Q) metalwork;
(R) plumbing and mechanical engineering installations;
(S) electrical installations;
(T) floor wall and ceiling finishings;
(U) glazing;
(V) painting and decorating;

(W) drainage;
(X) fencing.

Each work section is further divided into sub-headings. The purpose of these headings is to group work, materials and components of a similar nature together in the bill of quantities. This greatly assists the estimator in pricing the bill.

Use in practice

The SMM is not a simple document to use initially. A copy should be obtained by students, and by reference and class application gradually build up their ability to select the relevant information. In the taking-off process the SMM is used as follows:

1. to define the unit of measurement of the particular piece of work being measured;
2. to build up a suitable description of the work being measured which will enable the estimator to price the item.

Method of measurement of items of work not fully designed or to be the subject of post-contract selection

The SMM requires that work which cannot be measured or the extent of which is not known at the time the bill of quantities is prepared should be given as provisional sums.

Work to be undertaken by nominated sub-contractors and statutory undertakings and supplies made by nominated suppliers are to be given as prime cost sums. Provisional and prime cost sums were discussed in Chapter 2.

Chapter Eight

Measurement application

Introduction

We can now begin work on practical measurement exercises, putting into practice the basic principles studied earlier in the book. Level II is introductory and therefore the examples given are of a straightforward nature. Adjustments are kept to a minimum to assist in preparing simple bills of quantities.

Standard method of measurement

The use of the SMM is introduced by quoting the relevant clause reference numbers. Students may have difficulty in phrasing descriptions themselves initially and should concentrate on obtaining the correct dimensions. As they gain confidence and experience their ability in writing suitable descriptions will improve.

The measurement processes

We should at this stage remind ourselves of the four processes involved in preparing a bill of quantities:

1. taking-off;
2. squaring;
3. abstracting;
4. billing.

We must now apply these four processes to worked examples.

Taking-off

The drawings should be studied together with the specification, if given. Possibly the best approach to taking-off quantities in the early stages of study is to think how would I build this and then try and measure the quantities in that order. The student should write out a take-off list of items to be measured on a spare piece of paper or better still at the head of the first sheet of dimension paper relating to the example.

In the examples that follow the right-hand side of the dimension paper is used for explanation purposes. In practice of course both columns of the paper would be used for taking-off.

Again in practice the dimensions and descriptions would be handwritten, the description being abbreviated to save time. For clarity and to avoid misunderstanding the descriptions in this book are fully described and set in typescript.

Squaring

Remember the squaring up of dimensions is undertaken when the taking-off process is complete. Where the unit of billing is the metre and dimensions are being transferred to bill paper the final quantity should be rounded up or down to the nearest whole unit. Where adjustments (deductions) occur these have been calculated out on the dimension paper before being transferred to the bill. See Fig. 8.1.

Abstracting

The examples in this book are of a simple nature and we can therefore bill direct from the take-off stage without the need of an abstract. Abstracting is used on more complex examples to arrive at total quantities and bill order. An example of abstracting will be given at Level III.

Billing

Many of the worked examples in taking-off are taken through to billing stage to enable the student to identify the end product of the process. For simplicity, general items such as plant and protection are left off at this stage. They are incorporated, however, into the final worked example and are fully explained at that time.

Where this book is used to supplement lectures the author suggests that students be asked to look up rates from published sources (see Part 3) and use these to price up the bills of quantities they have measured and prepared. Many building students enjoy getting an answer, which is not always so apparent to them when they simply take-off quantities.

Direct billing

Before proceeding to the worked examples Fig. 8.1 should be studied. This shows the methods adopted in the worked examples for squaring and for making adjustments.

2.50		Excavate topsoil to be preserved	In this book to keep the examples simple
4.00	10.00	average 150 mm deep	deductions have always been placed after

the description and dimensions to which
they refer.
The adjusted quantity is placed under a line
drawn in the description column. The
calculation in this case being

1.00		*Deduct*	$10.00 - 2.10 = 7.90 \text{ m}^2$
2.10	2.10	ditto.	

As the quantities and descriptions are
removed to bill so the description should

7.90 m²

be lined through.
This avoids them being taken twice.

Remember also the final quantity is
rounded up or down to the nearest metre
when being billed.

5.00		Excavate foundation trench	No adjustment is required in this case so
0.60		exceeding 0.30 m wide	the quantities and description are removed
0.60	1.80	commencing at underside of	direct to the bill. Here there are two
		topsoil strip maximum depth	descriptions and the same quantity applies
7.00		not exceeding 1.00 m.	to both. As each item is removed the
0.60			descriptions are lined through. When all
0.60	2.52		items within the bracket are removed a

vertical line is drawn through.

&

		Remove excavated material	When all items have been lined through on
		from site.	the take-off the bill is complete.
	4.32		

Fig. 8.1 Squaring, adjusting and lining through for direct billing

Example 8.1.
Excavation to reduce levels

Figure 8.2 shows a plan and section of a building site which at present has a sloping surface. Before building work commences it is required that the surface be excavated to make the site level. The required reduced level is 100.00.

The quantity surveyor must know whether excavated material is to be retained or removed from site before the work can be measured. The specification details for this site read as follows:

Topsoil excavation   Topsoil is to be excavated and retained on site for future use. Assume average depth of topsoil to be 150 mm deep.

Sub-soil excavation   Sub-soil to be excavated and removed from site.

In the following worked example the student should first study the drawing followed by the taking-off and then the simple bill of quantities.

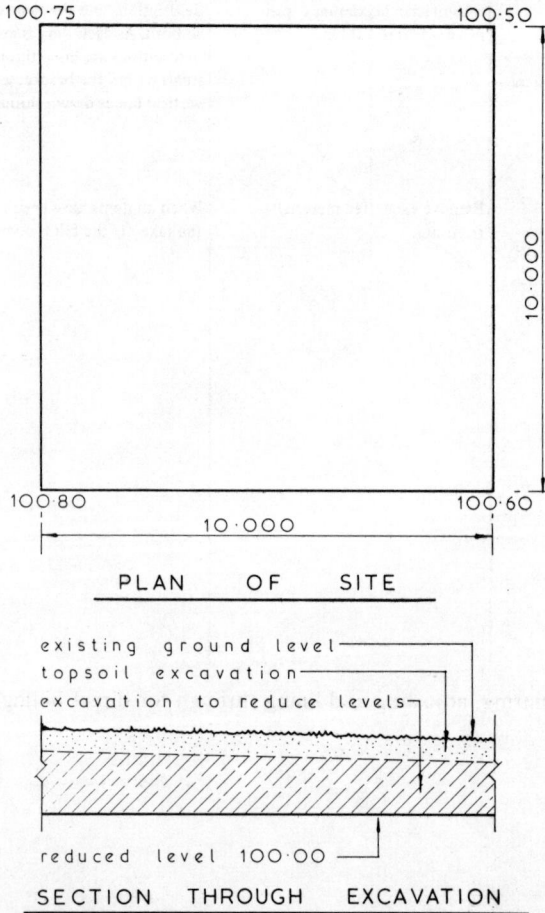

Fig. 8.2 Example 8.1: Excavation to reduce levels

Take-off for Fig. 8.2

EXCAVATION EXAMPLE

Take-off list

The take-off list is not part of the taking-off.

1) Excavate topsoil
2) Deposit topsoil
3) Excavate to reduce levels
4) Remove excavated material from site

Ground level		*Reduced level*	*Average depth of excavation*
100.75	←	100.00	0.75
100.50	←	100.00	0.50
100.80	–	100.00	0.80
100.60	–	100.00	0.60
sum of average depths			2.65

Waste calculation to find average depth of excavation. At each corner the reduced level required is deducted from the existing ground level. The sum of the depths is divided by 4 to find the average (0.662 m) depth of excavation. Topsoil depth of 150 mm is then deducted from 0.662 m to leave actual depth of reduced level excavation.

$\therefore$ average depth = $\dfrac{2.65}{4}$

= 0.662

Deduct topsoil 150 mm thick   0.150

$\therefore$ average depth of sub-soil
   excavation   =   0.512

10.00		Excavate topsoil to be
10.00	100.00	preserved average
———		150 mm deep.

SMM D.9
Measured in m^2 stating the average depth of excavation.

10.00		Deposit preserved topsoil
10.00		in temporary spoil heap
0.15	15.00	for re-use average 15 m
———		from excavation.

SMM D.31
Measured in m^3 stating average distance of spoil heap from excavation in metres or kilometres.

10.00		Excavate to reduce levels
10.00		maximum depth not
0.51	51.00	exceeding 1.00 m

SMM D.13.3
Measured in m^3 depth classified as clause D.11.

&

Remove excavated material from site

SMM D.29
Measured in m^3. Provision of a suitable tip by the contractor is deemed to be included.

The take-off can now be billed. The SMM work section sub-headings are used as a guide to bill order. The order of the bill will not be the same as the take-off.

Bill for Fig. 8.2

BILL OF QUANTITIES

EXCAVATION EXAMPLE

	Site preparation		
A	Excavate topsoil to be preserved average 150 mm deep.	100	m^2
	Excavation		
B	Excavate to reduce levels maximum depth not exceeding 1.00 m.	51	m^3
	Disposal of excavated material		
C	Remove surplus excavated material from site.	51	m^3
D	Deposit preserved topsoil in temporary spoil heaps for re-use average 15 m from excavation.	15	m^3

To collection £

Example 8.2.
Foundation bases and trenches

This type of foundation may not have been studied by students in technology during Level II. It is, however, a simple measurement exercise and is introduced to teach some of the basic principles of measurement relating to foundation work. A foundation design such as this is suitable for a framed building (pre-cast concrete or steel framed) where the bases support the weight of the frame and the foundation trenches the infill wall panels. Normally the method of securing the frame to the foundation bases (by leaving pockets in the concrete, or casting bolts into the concrete) would be measured. For simplicity we will ignore these details.

Reference should now be made to Figs. 8.3 and 8.4. Note that the site is flat and that topsoil has previously been removed.

Consider how the work would be carried out on site:

1. Excavate the foundation bases and trenches.
2. If necessary provide temporary support to the sides of the trenches. The SMM refers to this as earthwork support and requires it to be measured to the side of any excavation which exceeds 0.25 m in height (SMM D.15.a). Students may feel that earthwork support would not be necessary at this depth of excavation. The logic of the SMM is that any excavation exceeding 0.25 m deep that may require earthwork support should be measured. The contractor can elect not to price this item if in his judgement the ground conditions do not warrant support. Note that only the measurement of the excavation area requiring support is given, the method and materials used are entirely at the contractor's discretion.
3. Bottom up base of excavation including concrete level pegs.
4. Pour concrete into foundation.

This construction process is now expressed as a take-off list on the worked example that follows on page 110.

FOUNDATION   PLAN

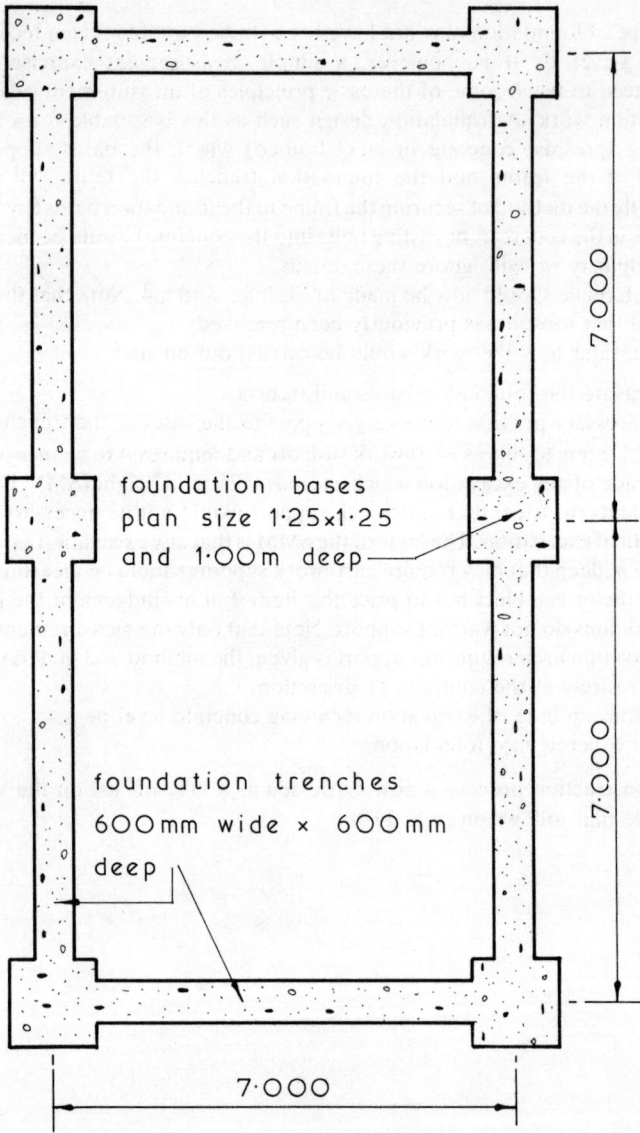

foundation bases
plan size 1·25x1·25
and 1·00m deep

foundation trenches
600mm wide × 600mm
deep

7·000

7·000

7·000

Fig. 8.3   Example 8.2: Foundation bases and trenches

structural frame

surface strip level
topsoil assumed previously removed

600

G.L.

400

1·250

plain concrete (1:3:6–19)
in foundations

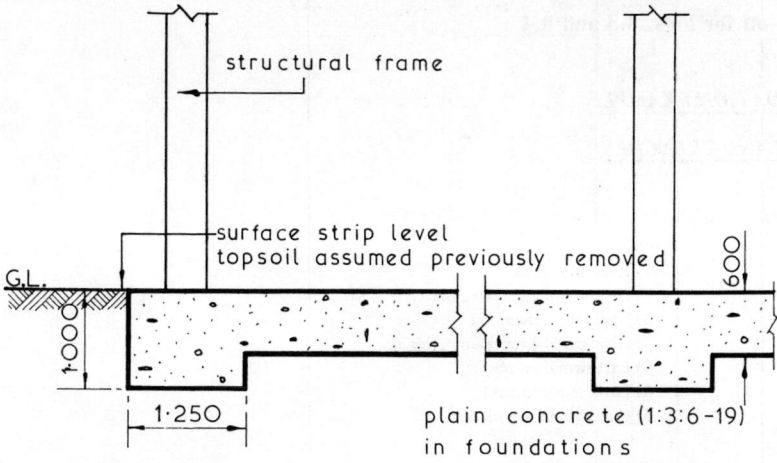

SECTION THROUGH FOUNDATION

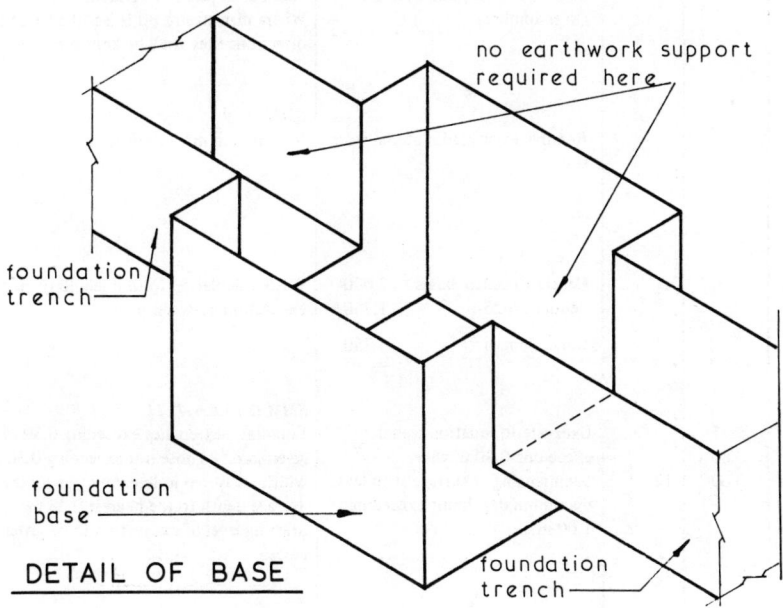

no earthwork support
required here

foundation
trench

foundation
base

DETAIL OF BASE

foundation
trench

Fig. 8.4 Example 8.2: Details of foundation bases

Take-off for Figs. 8.3 and 8.4

FOUNDATION EXAMPLE

BASES AND TRENCHES

Take-off list

1) Excavate bases
2) Disposal of excavated material
3) Excavate trenches
4) Disposal of excavated material
5) Earthwork support
6) Level and compact
7) Concrete in foundations

6/	1.25 1.25 1.00	9.38	Excavate pits to receive bases of stanchions commencing at surface strip level maximum depth not exceeding 1.00 m. (In 6 number)	*SMM D.13.5, D.11* Foundation bases are termed 'pits' in the SMM. Starting level of excavation and depth as clause D.11 to be given. Number of pits to be stated. Where plan size of pit is less than 1.25 in both directions they shall be kept separate.

&

			Remove excavated material from site.	*SMM D.29* Measured in m³ as before.

Centre to centre bases	7.000		Waste calculation to find length of trench excavation between pits.
deduct 2/625	1.250		
trench length	5.750		

6/	5.75 0.60 0.60	12.42	Excavate foundation trench exceeding 0.30 m wide commencing at surface strip level maximum depth not exceeding 1.00 m.	*SMM D.13.6.b, D.11* Foundation trenches exceeding 0.30 m wide given in m³. Those not exceeding 0.30 m in width measured in linear metres stating average depth to the nearest 0.25 m. Starting level of excavation to be given in both cases.

&

			Remove excavated material from site.	*SMM D.29*

Girth of pit	4/1.25	5.000	Waste calculation to find perimeter of pit for earthwork support.

FOUNDATION EXAMPLE continued

6/	5.00		Earthwork support maximum	
	1.00	30.00	depth not exceeding 1.00 m and	
			width between opposing faces not	
6/2/	5.75		exceeding 2.00 m.	
	0.60	41.40		
		71.40		

SMM D.15, D.16, D.17
Depth classified as for excavation D.11. Width between opposing faces given as clause D.17.
not exceeding 2.00 m
        2.00 m – 4.00 m
    exceeding 4.00 m
In this case both the pits and trenches are in the same depth and width classifications and may be grouped together.
Note 6 pits so × 6.
2 sides to each trench and 6 trenches.

			Deduct	
6/2/	0.60		ditto	
	0.60	4.32		
			(junction of trench and pit)	
		67.08 m²		

See Fig. 8.4. No earthwork support where trench excavation meets pit.
2 number per pit and 6 pits.
Wording in brackets called a 'signpost'.
Indicates thought process of taker-off, and is not carried to bill.
Note use of word 'ditto' to save writing description again.

6/	1.25		Level and compact	
	1.25	9.38	base of excavation to	
6/	5.75		receive concrete.	
	0.60	20.70		
		30.08		

SMM D.40
Measured in m².
Bottoming up of excavation prior to placement of concrete.

			In-situ concrete	
6/	1.25		Plain concrete in foundation	
	1.25		trenches (1 : 3 : 6 – 19) over	
	1.00	9.38	300 mm thick poured against	
6/	5.75		face of excavation.	
	0.60			
	0.60	12.42		
		21.80		

SMM F.6.2, F.5.2, F.46
Foundation trenches are deemed to include column and pier bases which are not isolated see clause F.6.2. Depth of concrete given as clause F.5.2.
not exceeding 100 mm
        100 – 150 mm
        150 – 300 mm
    exceeding 300 mm thick.

The take-off is now billed using the subheadings of the work section of the SMM to determine bill order.

Bill for Figs. 8.3 and 8.4

BILL OF QUANTITIES

FOUNDATION BASES AND TRENCHES

Excavation

A	Excavate pits to receive bases of stanchions commencing at surface strip level maximum depth not exceeding 1.00 m (in 6 number)	9	m³
B	Excavate foundation trench exceeding 0.30 m wide commencing at surface strip level maximum depth not exceeding 1.00 m.	12	m³

Earthwork support

C	Earthwork support maximum depth not exceeding 1.00 m and width between opposing faces not exceeding 2.00 m.	67	m³

Disposal of excavated material

D	Remove surplus excavated material from site.	22	m³

Surface treatments

E	Level and compact base of excavation to receive concrete.	30	m²

In-situ concrete

F	Concrete in foundation trenches (1 : 3 : 6 – 19) over 300 mm thick poured against faces of excavation.	22	m³

To collection   £

Example 8.3.
Brickwork and blockwork generally

Measurement of brickwork and blockwork

This section of the SMM is a large one. The student cannot hope to absorb all the particular clauses at Level II. By learning to apply a few basic rules, however, a large range of simple brickwork and blockwork can be measured. The student may then build on this basic knowledge in later years of study to measure more complicated work.

Terminology
Solid walls: For measurement purposes walls are described by their thickness as Fig. 8.5 illustrates.

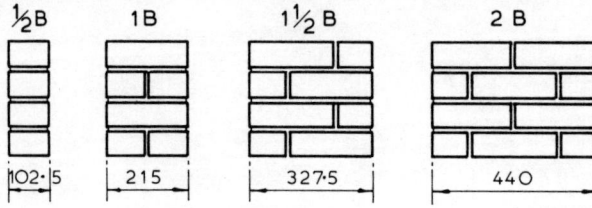

Fig. 8.5 Wall thicknesses

Cavity walls: Cavity walls are described as being in skins of hollow walls (SMM G.5.3.c). Forming the cavity between the two walls is measured separately in m² (SMM clause G.9). Figure 8.6 illustrates a cavity wall.

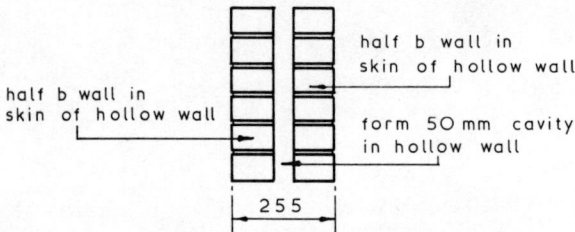

Fig. 8.6 Cavity walls

Window and door openings
Earlier in the book we saw that where window and door openings occur these would be measured through initially. Adjustments for these are made at a later stage in the take-off (dealt with in Level III). The drawings used in this book for simplicity do not show any window or door openings.

Classification of work
Clause G.3, SMM, requires that work shall be classified and billed under the following headings:

1. work in foundations;
2. work in load-bearing superstructures;
3. work in non-lead-bearing superstructures;

Categories of work
Clause G.5.3 (a–m), SMM, lists the different types of situation in which

brickwork is commonly used. Work being measured should always be described as being in one of these categories. At Level II we shall concern ourselves with just three:

G.5.3.a walls;
G.5.3.c skins of hollow walls;
G.5.4 projection of attached piers.

Measurement of brickwork
Brick walls may be built in common or facing bricks or a combination of both. Again some surfaces of walls are pointed whilst others are not. To enable brickwork to be measured in a form in which it can be priced the SMM categorizes brickwork in the following manner:

1. brickwork (SMM G.5.);
2. brick facework and fair face (SMM G.14.1, G.14.2, and G.14.3);
3. walls built entirely in facing bricks or fair faced both sides. This applies to half-brick and one-brick-thick walls only (SMM G.14.9).

If the student seeks to learn one thing about measuring brickwork at this level then it should be to identify the type of wall from a drawing and the method of measuring it.

The following examples are intended for this purpose:
Example 8.4 – Brickwork;
Example 8.5 – Brick facework and fair face;
Example 8.6 – Walls built entirely in facing bricks or fair faced both sides.

Example 8.4.
Brickwork

Brickwork in this sense means a wall of any thickness built of common bricks
which does not incorporate any facing bricks or pointed finish to the surface of
the wall. Walls of this type are often plastered when finished.

Brickwork should be measured the mean length × the average height
(SMM G.4.1). In practice the mean length is the centre-line length of the wall.

Figure 8.7. indicates a single-storey building having a flat roof supported
on a plain brick wall. The brickwork above d.p.c. level only is to be measured.

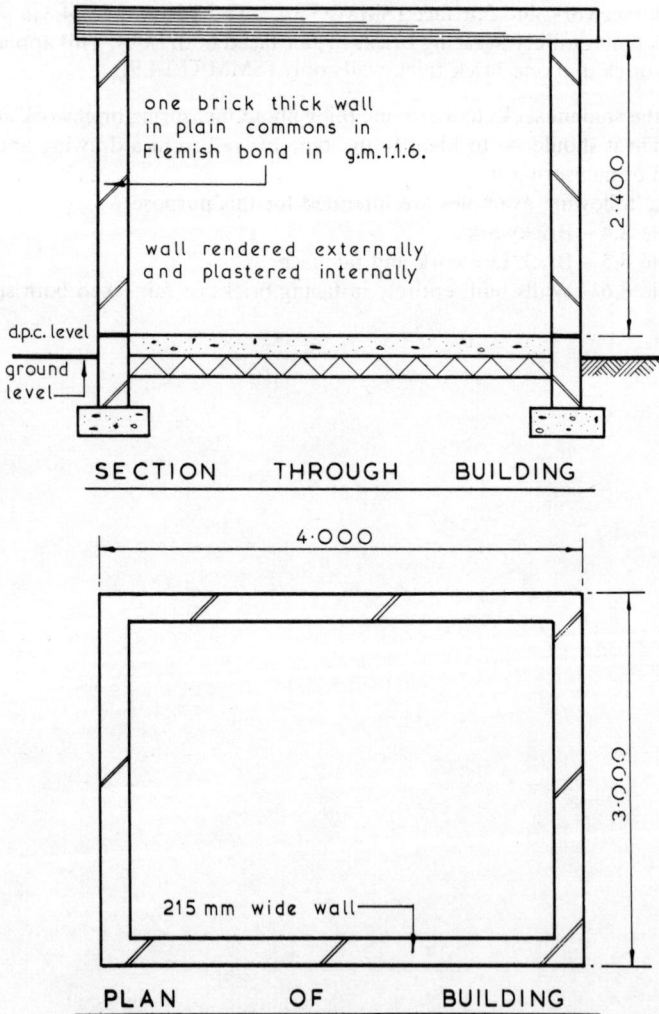

SECTION THROUGH BUILDING

PLAN OF BUILDING

Fig. 8.7 Example 8.4: Brickwork

Take-off for Fig. 8.7

BRICKWORK

The following in

load bearing superstructure

work above d.p.c. level

centre-line of wall | Classification as SMM G.3.1.

2/4.000	8.000	
2/3.000	6.000	

Waste calculation to find length of centre-line of brickwork.

external perimeter	14.000	
deduct corners		
4/2/107.5	0.860	
	13.140	

Deduction of 2 × half wall thickness per corner.

Brickwork

Sub-headings used as SMM to assist in logical take-off.

13.14	One brick thick wall in 65 mm
2.40	plain commons in Flemish bond in gauged mortar 1.1.6

SMM G.5
Measured in m² stating kind, quality and size of bricks.
Type of bond and composition and mix of mortar.

Internal perimeter

external perimeter	14.000	
deduct corners		
4/2/215	1.720	
	12.280	

Waste calculation to build up internal perimeter length.
Note 2 × whole wall width deducted per corner.

12.28	Rake out joints to form key for
2.40	plaster

SMM G.40
Internal face of wall to be plastered. Raking out joints measured in m² stating purpose for which key is required.

14.00	Ditto for render.
2.40	

SMM G.40
Similar item to above, but for rendering to external face of wall.

Example 8.5.
Brick facework and fair face

Brick facework

Many solid walls are built in a combination of common and facing bricks. Facing bricks are used on the exposed face of the wall where a pleasing finish is required. Where the bricks cannot be seen in the body of the wall or where they are to be plastered over then common bricks are used because they are cheaper than good-quality facing bricks.

The use of facing bricks in combination with common bricks is called brick facework and can be to one or both sides of a wall as Fig. 8.8 shows. Brick facework should be measured as follows:

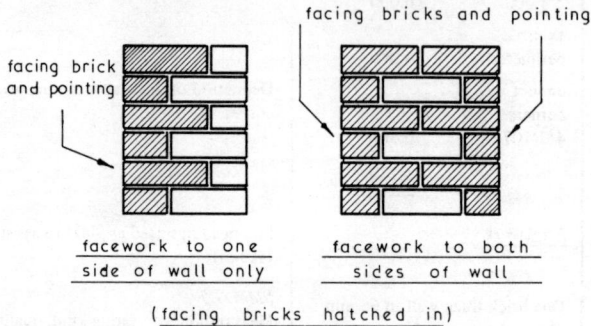

Fig. 8.8 Brick facework

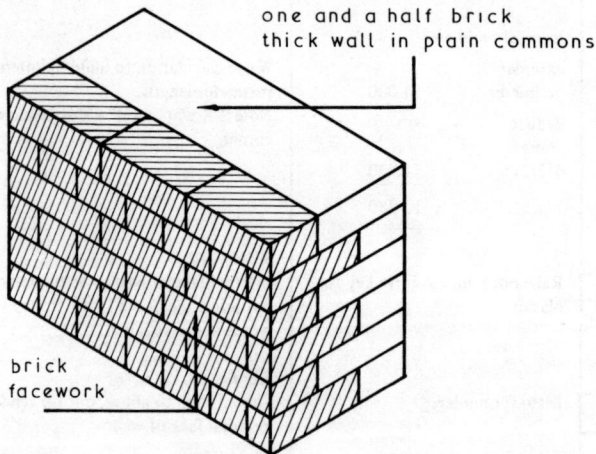

Fig. 8.9 Brick facework

1. measure the wall as common brickwork in m² using centre-line length × vertical height.
2. measure area of facework in m² on the exposed surface of the wall × vertical height as 'extra over' common brickwork for facework describing the type of facing bricks, bond, mortar if different for that used for the common bricks and method of pointing required. The term 'extra over' means the extra cost of facing bricks compared to common bricks plus the

bricklayer's time in laying and pointing. To understand this more fully consider Fig. 8.9.

This wall would be measured and priced as follows:

Measured description	*Estimator's method of price build-up*
1. One and a half brick thick wall in plain commons in English bond pointed mortar 1:1:6.	Assume estimator calculates a price of £18 per m² for this item.
2. Extra over common brickwork for facework in Wealden stocks in English bond pointed with a neat rubbed-in joint.	The estimator knows that English bond requires 89 facing bricks per m². Extra cost of materials equals difference in cost of 89 common and facing bricks, say £5. The extra cost of labour is calculated at say £2 per m² for the bricklayer's extra time in laying and pointing the facework. Total extra over cost then is £5 + £2 = £7.

Fair face finish on brickwork
This simply means a wall built in plain common bricks, the surface of which is built in a neat manner with the mortar joints being pointed. This differs from facework where facing bricks and pointing were used to the surface of the wall. Fair face gives an economical finish to a wall built in plain bricks. Paint can be applied direct to the wall as a finish and no plaster is then used.

The extra cost of a fair face finish to a wall is the bricklayer's extra time spent in building the surface of the wall true and in pointing up the mortar joints. As with facework, fair face can be to one or both sides of a wall as in Fig. 8.10.

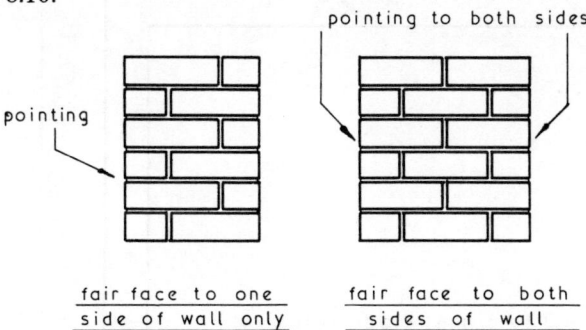

Fig. 8.10 Fair face

Fair face finish on brickwork is measured in a similar manner to facework:
1. measure the wall in m² as previously described for plain brickwork stating thickness, type of bricks, type of bond and mortar mix;
2. measure the area of fair face on the exposed surface of the wall in m² as

120   extra over common brickwork for a fair face finish stating the method of
pointing.

Figure 8.11 indicates a building having walls built in facework and a fair
face finish and the student should work carefully through the dimensions
relating to the drawing.

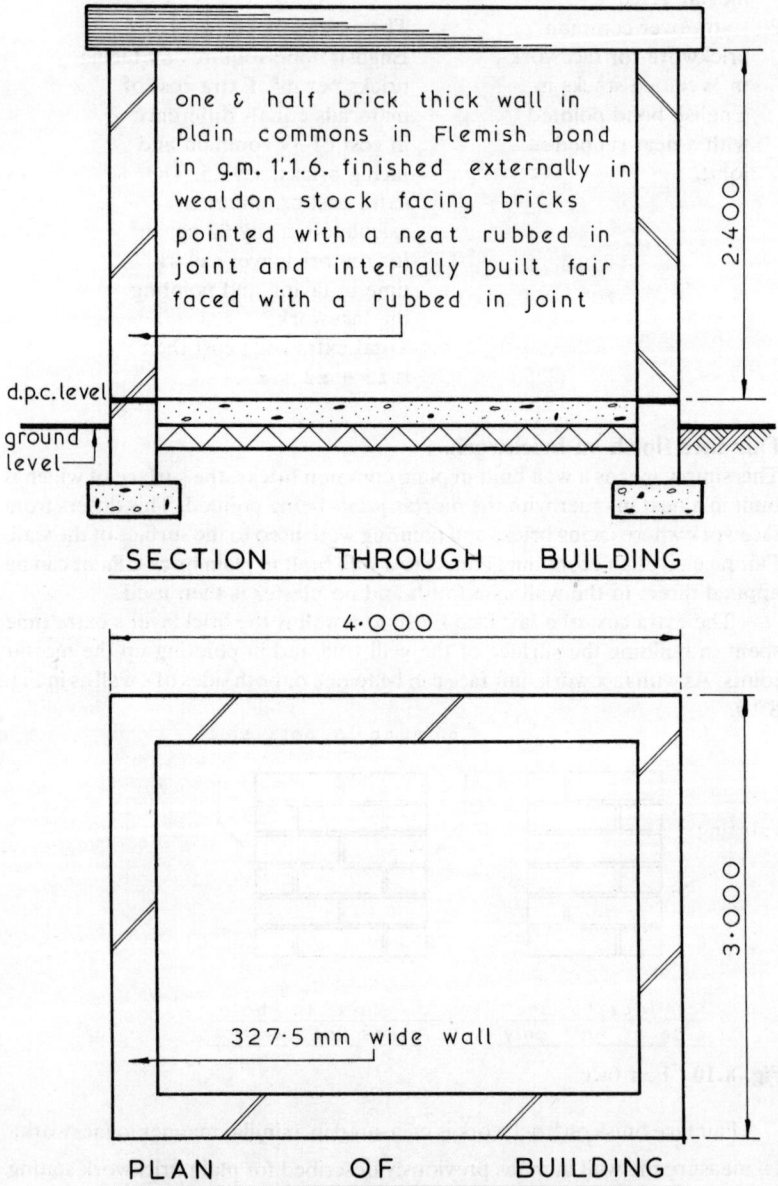

Fig. 8.11   Example 8.5: Brick facework and fair face

Take-off for Fig. 8.11

BRICK FACEWORK AND FAIR FACE

		The following in	Classification as SMM G.3.1.
		load bearing superstructure	

		work above d.p.c. level	
		centre-line of wall	Waste calculation to find length of centre-line of brickwork.
		2/4.000 8.000	
		2/3.000 6.000	
		external	
		perimeter 14.000	
		deduct	Deduction of 2 × half wall thickness per corner.
		corners	
		4/2/163.75 1.310	
		12.690	

		Brickwork	Sub-headings used as SMM to assist in logical take-off.
			SMM G.5
12.69		One and a half brick thick wall in	Plain brickwork as before.
2.40		plain commons in Flemish bond	
		in gauged mortar 1.1.6	

		Brick facework and fair face	Sub-headings from SMM
			SMM G.14.1, G.14.2, G.14.3
14.00		Extra over common brickwork	Measured in m² as extra over on exposed face of wall. Following should be stated.
2.40		for facework in Wealden stock	Kind and quality of facing bricks. Size of bricks where different from those in the body of the work.
		facing bricks in Flemish bond	
		pointed with a neat rubbed in joint	Type of bond. Mortar mix stated only if different from that used in the body of the work.
			The method of pointing.

		Internal perimeter	Waste calculation to find length of internal face of wall.
		external	
		perimeter 14.000	
		deduct	Deduction of 2 × wall thickness per corner.
		corners	
		4/2/327.5 2.620	
		11.380	

			SMM G.14.1, G.14.2, G.14.3
11.38		Extra over common brickwork	Measured in m² as extra over on exposed face of wall.
2.40		for fair face finish pointed with	Details of pointing given in description.
		a neat rubbed in joint	

Example 8.6.
Walls built entirely in facing brick or built fair faced both sides

Clause G.14.9 of the SMM states that walls built with a fair face to both sides or entirely of facing bricks that are a *half brick or one brick thick* should be measured separately. Such walls are measured their mean length × the average height.

A good example of a wall built entirely in facing bricks would be the external skin of a cavity wall.

Figure 8.12 details a building which includes walls which would be measured under this category. It also introduces the measurement of cavity wall construction.

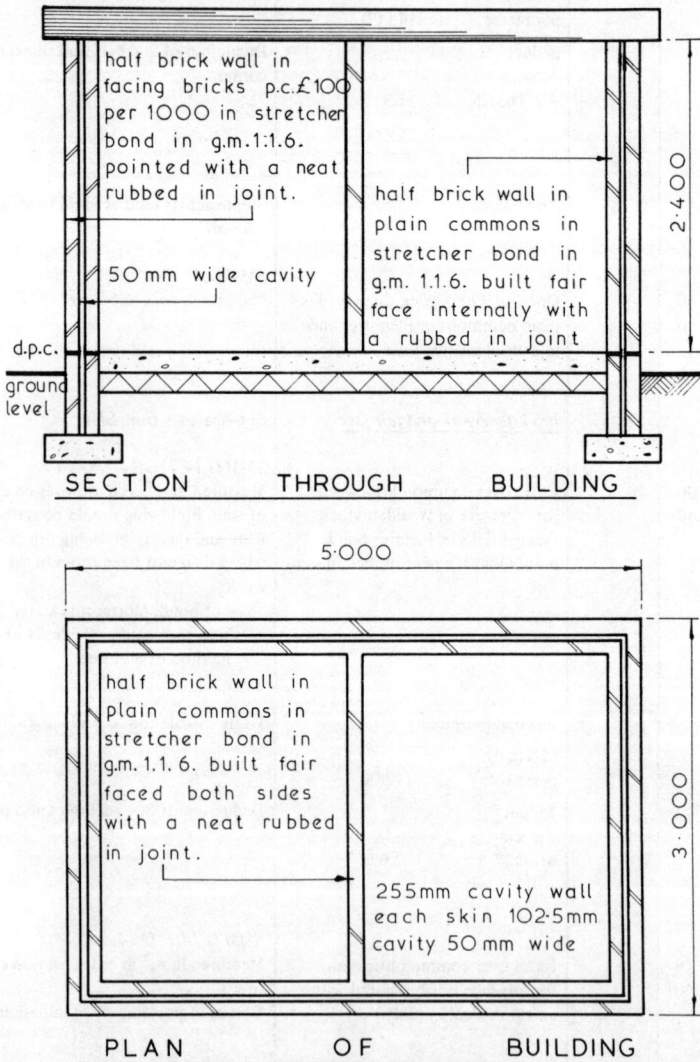

Fig. 8.12 Example 8.6: Walls entirely in facing bricks or fair faced both sides

WALLS ENTIRELY IN FACING BRICKS OR FAIR FACED BOTH SIDES

		Take-off list	Although this example is primarily to consider walls built entirely in facing bricks or fair faced both sides, the other categories are also to be found.
		Brickwork	
		½B internal skin of wall.	
		Form cavity.	By producing a take-off list using the three categories of brickwork each wall on the drawing is examined and noted under its relevant clause heading. This makes for simpler taking-off.
		Brick facework	
		Fair face finish to internal wall.	
		Walls entirely in facing bricks or fair faced both sides	
		½B external skin.	
		½B party wall.	

The following in

load bearing superstructure

Classification as SMM G.3.1.

Work above d.p.c. level
centre-line of internal skin

Waste calculation to find length of centre-line of internal skin of wall.

2/5.000	10.000
2/3.000	6.000

external	
perimeter	16.000

deduct corners

Deduction

102.5	102.5 external skin
50	50 width of cavity
51.25	51.25 half of internal skin
203.75	203.75 × 2 is deduction per corner.

4/2/203.75	1.630
	14.370

Brickwork

SMM G.5, G.5.3.c
Cavity walls described as in skins of hollow walls clause G.5.3.c.

14.37	Half brick thick wall in skin of
2.40	hollow wall in 65 mm plain commons in stretcher bond in gauged mortar 1.1.6.

centre-line of cavity

Waste calculation to find length of centre line of cavity.

external	
perimeter	16.000

deduct corners

Deduction is:

102.5	102.5 external skin
25	25 half of cavity width
127.5	127.5 × 2 per corner

4/2/127.5	1.020
	14.980

SMM G.9.1

14.98	Form 50 mm wide cavity in
2.40	hollow wall including 4 no. galvanized butterfly wall ties per m².

Measured in m² stating width of cavity. Wall ties to be given in the description stating type and spacing.

		Brick facework and fair face	Sub-heading
		external perimeter 16.000	Waste calculation to find length of fair face work.
		deduct corners 4/2/255 2.040	To find internal perimeter deduct 2 X total wall thickness per corner from external perimeter.
		13.960	Dividing wall junction with external wall must also be deducted, no fair face at that point.
		deduct junction of party walls with external wall 2/102.5 0.205	
		13.755	
13.76		Extra over common brickwork for fair face finish pointed with a rubbed in joint.	*SMM G.14.1, G.14.2, G.14.3* Measured on exposed face in m² as extra over common brickwork.
2.40			
		Walls entirely in facing bricks or fair faced both sides	Sub-heading
		centre-line external skin external perimeter 16.000	Waste calculation to find centre-line of external skins of wall.
		deduct corners 4/2/51.25 0.410	Deduct ½ external wall thickness X 2 per corner.
		15.590	
15.59		Half brick thick wall in skin of hollow wall entirely in facing bricks (prime cost £100 per 1000) in stretcher bond in gauged mortar 1.1.6 pointed with a neat rubbed in joint	*SMM G.14.9, G.14.2* Measured on centre-line. Note use of prime cost sum for bricks rather than specifying type of brick.
2.40			
		dividing wall length 3.000	Waste calculation to find length of dividing wall.
		deduct 2/255 0.510	
		2.490	
2.49		Half brick thick wall in 65 mm plain commons in stretcher bond in gauged mortar 1.1.6 built fair both sides with a neat rubbed in joint	*SMM G.14.9* Because this wall is to be built fair faced both sides it is necessary to mention this in the description.
2.40			

Example 8.7.
Brickwork projections

At Level II the most common projection found in brickwork is an attached pier. Clause G.54 indicates that attached piers are measured in linear metres by giving the height as the dimension and stating width and depth in the description. This may be explained by referring to Fig. 8.13. This would be measured as in Fig. 8.13 and take-off.

Projections of footings and chimney breasts are kept separate, SMM G.5.3.e, and measured in m² stating the thickness of the projection measured beyond the face of the wall.

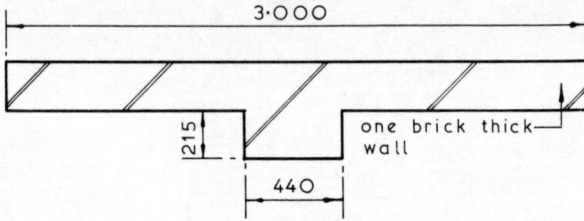

PLAN OF ATTACHED PIER

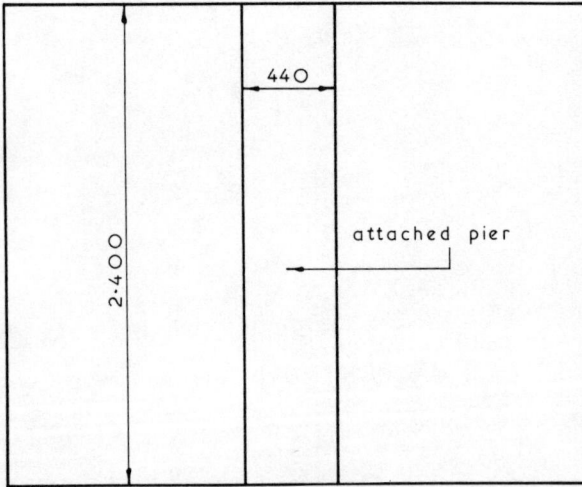

ELEVATION OF ATTACHED PIER

Fig. 8.13 Example 8.7: Brickwork projections

Take-off for Fig. 8.13

BRICKWORK PROJECTIONS

3.00 2.40	*Brickwork* One brick wall in plain commons in Flemish bond in gauged mortar 1 : 1 : 6	*SMM G.5, G.5.3* The one brick thick wall is measured as previously described.
2.40	Projection in attached pier 440 wide × 215 projection	*SMM G.5.4* Height of attached pier given as 2.40 m and depth and width stated in the description. Note the projection is measured beyond the face of the wall.

Example 8.8
Blockwork

As with brickwork, once the student has learnt a few rules blockwork of a simple nature can be measured.

The main rules to learn are as follows:

1. blockwork shall be measured in m² stating the thickness, SMM G.27;
2. classification of blockwork shall be as SMM G.27.1a–h inclusive. At this level the two common classifications are in:
   (a) walls,
   (b) skins of hollow walls;
3. where blockwork is required to have a fair face finish this shall be given in the description of the general blockwork stating whether to one or both sides, SMM G.27.2;
4. bonding ends of new blockwork to other construction shall be measured in metres stating the thickness of the blockwork, SMM G.33.

Figure 8.14 indicates a plan of part of a building. The internal partitions are of 100 mm thick blockwork. As with brickwork where doorways and window openings occur the blockwork is measured straight through and the adjustments to the dimensions would be made at a later stage.

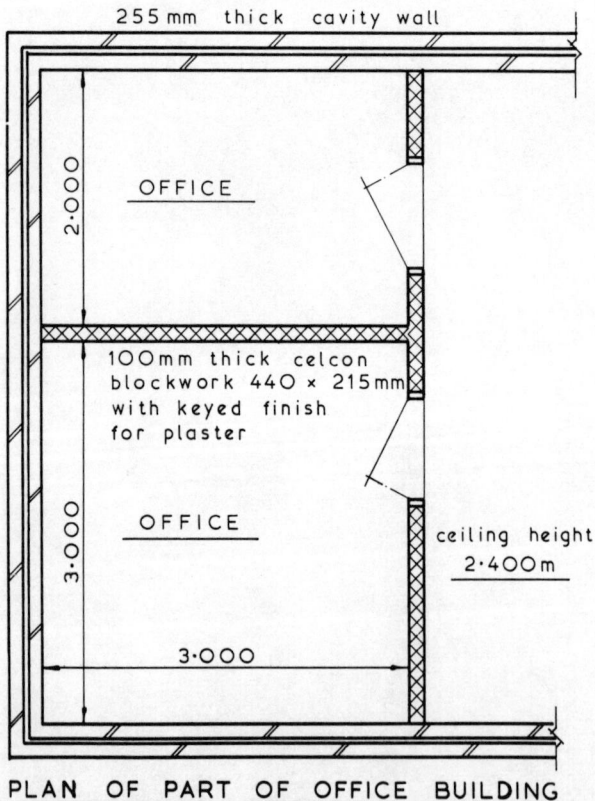

PLAN OF PART OF OFFICE BUILDING

Fig. 8.14 Example 8.8: Blockwork

Take-off for Fig. 8.14

BLOCKWORK EXAMPLE

		The following in		Classified as SMM G.3.

non-load bearing superstructure

		Office 1.	2.000	Length of blockwork partitions built up as
		partition	0.100	waste calculation.
		office 2.	3.000	Note no deductions are made for door openings
		dividing partition	3.000	at this stage.
			8.100	They are adjusted when the actual doors are measured.

	8.10	100 mm thick solid Celcon block-	*SMM G.26.1, G.27.1*
	2.40	work walls and partitions size	Blockwork is measured in m^2. Here the

100 mm thick solid Celcon blockwork walls and partitions size 440 × 215 mm with keyed finish bedded in gauged mortar 1.1.6

SMM G.26.1, G.27.1
Blockwork is measured in m^2. Here the manufacturer's name has been given. Alternatively the relevant BS number could be stated. The requirement of SMM G.26.e to state method of pointing is not mentioned because the wall will be plastered.

3/	2.40	Bonding ends of 100 mm thick blockwork to brickwork including forming pockets 102 mm deep to each alternate 3 courses of brickwork and extra material required.	*SMM G.33* Measured in linear metres. Note where bonding new blockwork to new brickwork forming pockets would be measured. Where bonding new blockwork to existing brickwork cutting pockets is measured. Extra material for bonding given in the description.

Example 8.9.
Boundary wall

This example enables further brickwork to be measured, together with foundation work and pre-cast concrete items.

The foundation detailed on the drawing Fig. 8.15 is of the trench fill type of construction. Trench fill using mass concrete is perhaps not so common as the more traditional strip foundation. They are simpler to measure, however, as no adjustment for backfill is required. The author bases exercises on trench fill methods at Level II and introduces traditional strip foundations during Level III by which time students have gained some basic experience in the subject and can better understand the adjustments required.

Figure 8.15 should be studied together with Figs. 8.16 and 8.17 which by means of diagrams indicate the stages of measurement up to d.p.c. level. In both the take-off and the subsequent bill, work below d.p.c. level is kept separate from work in superstructures as is the normal practice.

Specification details

Topsoil excavation	To be preserved on site for later use.
Trench excavation	Excavated material to be removed from site.
Foundation concrete	Plain mass concrete (1:3:6 – 19).
Foundation brickwork	In calcium silicate bricks class 3 in Flemish bond in cement mortar (1:4)
Brickwork in superstructure	Above d.p.c. level in plain commons in Flemish bond in gauged mortar (1:1:6). Facework in Ockley purple multi-facing bricks in Flemish bond in gauged mortar 1:1:6 pointed with a neat rubbed-in joint. Facework extends to three courses below d.p.c. level.
Damp-proof courses	Hessian-based bituminous felt to BS 743.
Coping stone	Pre-cast concrete to BS 3798 size 427 × 150 mm deep.

precast concrete
saddleback coping
to b.s. 3798

one and a half thick
brick wall in plain
commons in Flemish
bond in gauged
mortar (1·1·6).
finished externally
in brick facework
in ockley purple multi
facing bricks pointed
with a neat rubbed
in joint.

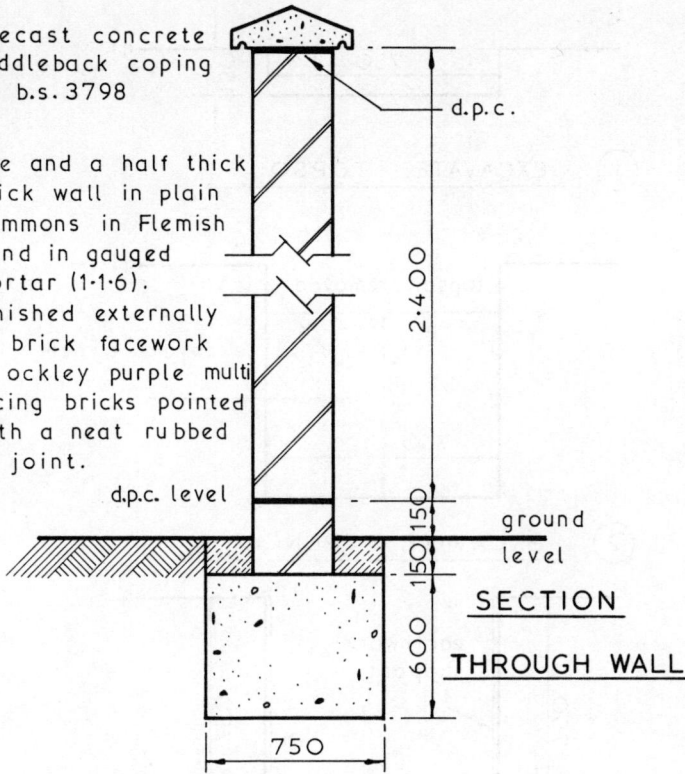

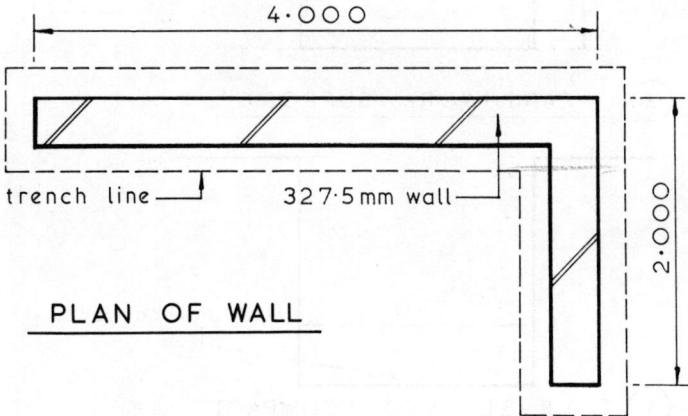

d.p.c. level

d.p.c.

2·400

150 150

600

ground
level

750

SECTION

THROUGH WALL

4·000

600

trench line

327·5 mm wall

2·000

PLAN OF WALL

Fig. 8.15 Example 8.9: Boundary wall

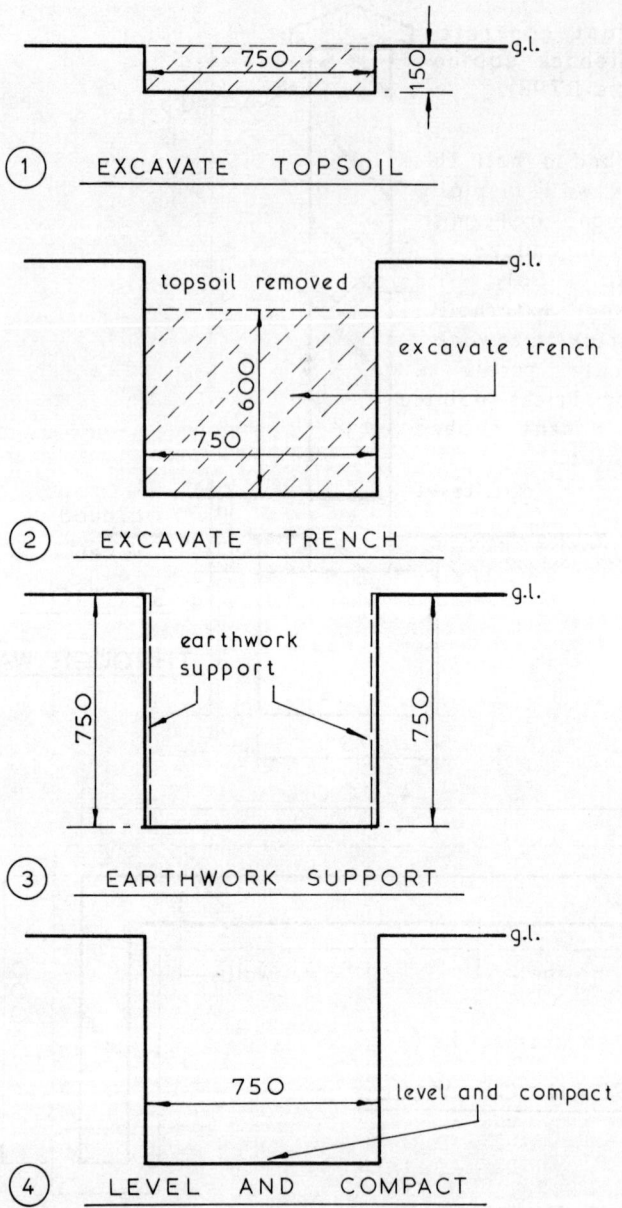

Fig. 8.16 Example 8.9: Measurement sequence

(5) CONCRETE IN FOUNDATION

(6) BRICKWORK UPTO D.P.C. LEVEL

(7) EARTH FILLING WITH TOPSOIL

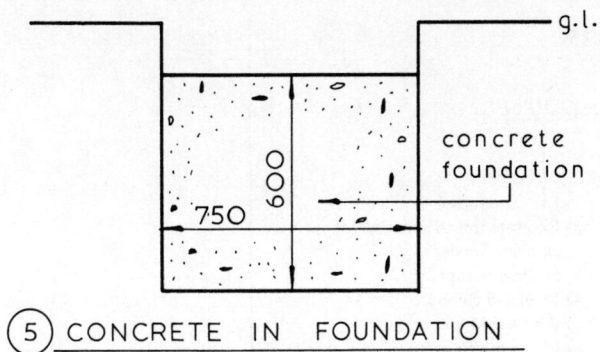

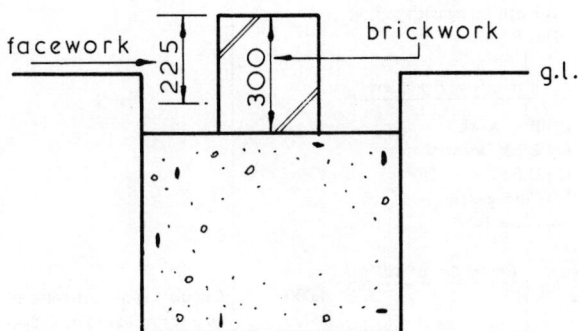

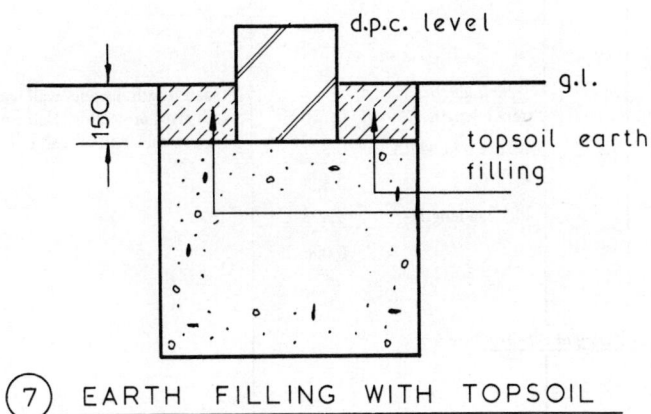

Fig. 8.17 Example 8.9: Measurement sequence

Take-off for Fig. 8.15

BOUNDARY WALL EXAMPLE

Take-off list

1) Excavate topsoil and deposit
2) Excavate trench and disposal of
   excavated material
3) Earthwork support
4) Level and compact
5) Concrete in foundation
6) Brickwork to damp-proof course level
7) Brick facework
8) Earth filling with topsoil
9) D.P.C.

Work above d.p.c. level

10) Brickwork
11) Brick facework
12) D.P.C.
13) Coping stone

Centre-line of wall	
	4.000
	2.000
	6.000
deduct corner	
2/163.75	0.327
	5.673

Calculation of centre-line of wall.
One corner has to be deducted.

Trench length	
trench length	0.750
deduct wall width	0.327
	0.423
add wall length	5.673
	6.096

Trench length. Deduct wall width from trench width gives projection. Half projection must be added to each end of wall to give trench length.

The following in sub-structure work

up to d.p.c. level

6.10		Excavate topsoil to be preserved
0.75	4.58	average 150 mm deep

SMM D.9
Measured in m² stating average depth.

6.10		Deposit preserved topsoil in
0.75		temporary spoil heap for re-use
0.15	0.69	average 15 m from excavation

SMM D.31
Topsoil to be preserved measured in m³ stating distance of spoil heap from excavation.

BOUNDARY WALL EXAMPLE continued

	6.10		Excavate foundation trench	*SMM D.6.b, D.11*
	0.75		exceeding 0.30 m wide	Measured in m³. State starting level of
	0.60	2.75	commencing at surface strip level	excavation and depth range as D.11.
			maximum depth not exceeding	
			1.00 m	
			&	
				SMM D.29
			remove excavated material from	Provision of tip deemed included unless other-
			site	wise stated.
				Note use of ampersand. Measured in m³.
				SMM D.14, D.15, D.17
2/	6.10		Earthwork support maximum	Measured in m² stating maximum depth as
	0.75	9.15	depth not exceeding 1.00 m and	clause D.11 and width between opposing faces
			width between opposing faces not	as D.17.
2/	0.75		exceeding 2.00 m.	Note centre-line of trench × 2 gives length of
	0.75	1.13		both sides of trench. Ends of trench must also
		10.28		be measured.
				SMM D.40
	6.10		Level and compact base of	Measured in m².
	0.75	4.58	excavation to receive concrete	
				SMM F.6.2, F.5.2, F.4.6
	6.10		Plain concrete in foundations	Measured in m³ stating thickness as clause
	0.75		(1 : 3 : 6 – 19) over 300 mm thick	F.5.2.
	0.60	2.75	poured against face of excavation	
				SMM G.5
	5.67		One and a half thick brick wall in	Centre-line of wall × height.
	0.30	1.70	calcium silicate bricks class 3 in	State type of bricks, bond and mortar mix.
			Flemish bond in cement mortar	
			(1 : 4)	
				SMM G.14.1, G.14.2, G.14.3
2/	5.67		Extra over common brickwork	Facework measured as extra over on exposed
	0.23	2.61	for facework in Ockley purple	face × height.
			multi-facing bricks in Flemish	Note ends of wall must also be taken.
2/	0.33		bond pointed with a neat rubbed-	On level site facework is measured to one
	0.23	0.15	in joint	course below ground level. This gives height
		2.76		of 3 courses to d.p.c. level. On sloping sites
				height of facework must be calculated.
				SMM D.33, D.35
	6.10		Earth filling around excavations	Measured in m³. The volume of topsoil is first
	0.75		with topsoil from temporary spoil	measured. The volume of the wall is then
	0.15	0.67	heap	deducted. This leaves the correct volume of
				topsoil required as earth filling.
	5.67		*Deduct*	
	0.33		ditto	
	0.15	0.17	(wall)	
		0.50		

	5.67		Hessian based bituminous felt	*SMM G.37*
	0.33	1.87	d.p.c. to B.S. 743 in single	Damp-proof courses over 225 mm wide
			horizontal layer bedded in cement	measured in m². Under 225 mm wide
			mortar (1 : 4) with minimum	measured in metres stating width in description
			100 mm laps (measured net no	
			allowance for laps)	

Work above d.p.c. level

The following brickwork in non-load

bearing superstructure

	5.67		One and a half thick brick wall	*SMM G.5*
	2.40	13.61	in plain commons in Flemish	
			bond in gauged mortar (1 : 1 : 6)	
				SMM G.14
2/	5.67		Extra over common brickwork	
	2.40	27.22	for facework in Ockley purple	
			multi-facing bricks in Flemish	
2/	0.33		bond pointed with a neat	
	2.40	1.58	rubbed in joint	
		28.80		

	5.67		Hessian based bituminous felt	*SMM G.37*
	0.33	1.87	d.p.c. to B.S. 743 as before	Similar to damp-proof course measured
			described.	earlier in work up to d.p.c. level. No
				need to write description fully again.

Coping-stone			
		5.670	Waste calculation to find length of coping.
add overhang 2/50		0.100	Allow 50 mm overhang at each end of wall.
		5.770	

5.77	Pre-cast concrete saddleback	*SMM F.19.1 and 2*
	coping 427 × 150 mm deep to	Standard or stock pattern coping measured
	B.S. 3798 end jointed to form	in metres stating the size and catalogue or
	continuous run bedded and	reference number as applicable.
	pointed in gauged mortar	Purpose-made copings are also measured in
	(1 : 1 : 6)	metres, but where they cannot be adequately
		described shall be accompanied by a bill
		diagram showing a profile of the section
		required.

1	Angle to ditto	*SMM F.19.1 and 2*
		Angles are enumerated separately.

2	Fair end to ditto	*SMM F.19.1 and 2*
		Fair ends are enumerated separately.

The take-off may now be billed.

Bill for Fig. 8.15

BILL OF QUANTITIES

BOUNDARY WALL

SUB-STRUCTURE WORK

Site preparation

A	Excavate topsoil to be preserved average 150 mm deep.	5	m²				

Excavation

B	Excavate foundation trench commencing at surface strip level maximum depth not exceeding 1.00 m.	3	m³				

Earthwork support

C	Earthwork support maximum depth not exceeding 1.00 m and width between opposing faces not exceeding 2.00 m.	10	m²				

Disposal of excavated material

D	Remove surplus excavated material from site.	3	m³				
E	Deposit preserved topsoil in temporary spoil heap for re-use average 15 m from excavation.	1	m³				

Filling

F	Earth filling to excavation with topsoil from temporary spoil heaps.	1	m³				

Surface treatments

G	Level and compact base of excavation to receive concrete.	5	m²				

CONCRETE WORK

In-situ concrete

H	Concrete in foundation trenches (1 : 3 : 6 – 19) over 300 mm thick poured against faces of excavation.	3	m³				

BRICKWORK AND BLOCKWORK
Brickwork

I	One and a half thick brick wall calcium silicate bricks class 3 in Flemish bond in cement mortar (1 : 4)	2	m²				

		To collection			£		

BOUNDARY WALL continued
Brick facework

A | Extra over common brickwork for facework in Ockley purple multi-facing bricks in Flemish bond pointed with a neat rubbed in joint. | 3 | m²

Damp proof courses

B | Hessian based bituminous felt damp proof course to B.S. 743 in single horizontal layer bedded in cement mortar (1 : 4) with minimum 100 mm laps (measured nett no allowance for laps) | 2 | m²

THE FOLLOWING WORK IN SUPERSTRUCTURES

CONCRETE WORK

Precast concrete

C | Precast concrete saddleback coping 427 × 150 mm deep to B.S. 3798 end jointed to form continuous run bedded and pointed in gauged mortar (1 : 1 : 6) | 6 | m

D | Angles to ditto | 1 | nr.

E | Fair ends to ditto | 2 | nr.

BRICKWORK AND BLOCKWORK

In non-load bearing superstructures

Brickwork

F | One and a half thick brick wall in plain commons in Flemish bond in gauged mortar (1 : 1 : 6) | 14 | m²

Brick facework

G | Extra over common brickwork for facework in Ockley purple multi-facing bricks in Flemish bond pointed with a neat rubbed in joint. | 29 | m²

Damp proof courses

H | Hessian based bituminous felt damp proof course to B.S. 743 in single horizontal layer bedded in cement mortar (1 : 4) with minimum 100 mm laps (measured net no allowance for laps) | 2 | m²

To collection £

Example 8.10
Simple foundation

The value of measuring a simple foundation cannot be over-emphasized to the student. Three important sections of the SMM are involved, namely excavation and earthwork, concrete work and brickwork and blockwork. Greater use will also be made of centre-line calculations in this example and the student will realize why so much time was spent on learning to calculate these in Chapter 6 of this book.

Two alternative methods of measuring the brickwork are given because some students find this particularly difficult.

No bill of quantities is included after the take-off because it is very similar to Worked Example 8.14 at the end of this chapter.

Figure 8.18 details the foundation work which is again of trench fill construction. Reference should also be made to Fig. 8.19, 8.20 and 8.21 which show the sequence of measurement in diagram form.

Specification notes

Topsoil excavation	To be retained on site for later use.
Trench excavation	Excavated material to be removed from site.
Foundation concrete	Plain mass concrete (1:3:6 – 19)
Brickwork	102.5 mm brickwork in cavity construction in calcium silicate bricks class 3 in stretcher bond in cement mortar (1:4) Facing bricks measured to three courses below d.p.c. level. Allow the prime cost sum of £100 per 1000. Pointed with a rubbed-in joint.
Cavity	Form 50 mm wide cavity between skins of brickwork and fill with weak concrete to within 150 mm of d.p.c. level.
Oversite construction	
Hardcore	Clean brick hardcore 150 mm thick well compacted and consolidated and blinded with sand.
Damp-proof membrane	1000 gauge polythene with minimum 200 mm laps laid on blinded hardcore bed.
Concrete bed	Reinforced concrete bed (1:2:4 – 20) 150 mm thick incorporating fabric reinforcement to BS 4483 type B.196 weighing 3.05 kg/m² with minimum 150 mm end and side laps.
Damp-proof courses	Hessian-based bituminous felt to BS 743.

facing bricks

d.p.c. level

ground level

150

150

150

earth filling

255

concrete bed

fabric reinforcement

damp proof membrane

sand blinding

hardcore bed

750

mass concrete foundation

SECTION THROUGH FOUNDATION

600

trench line

600 | 255

PLAN OF

FOUNDATION

2·500

4·000

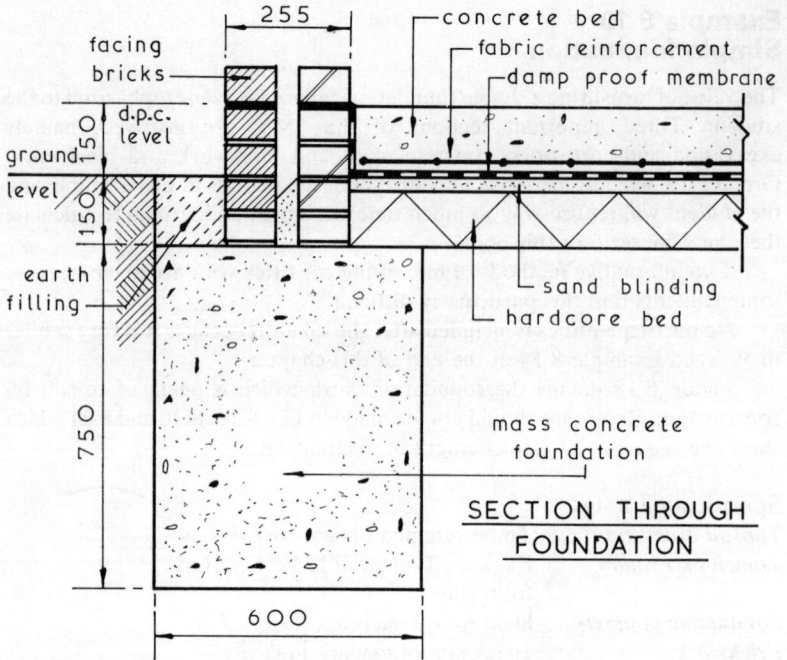

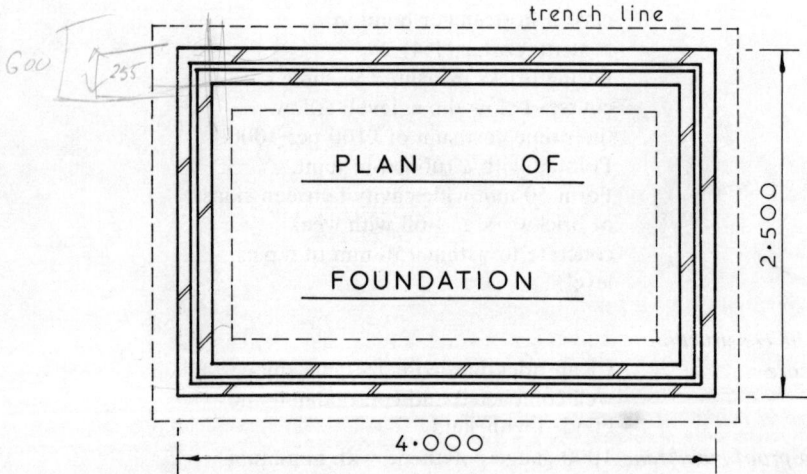

Fig. 8.18 Example 8.10: Simple foundation

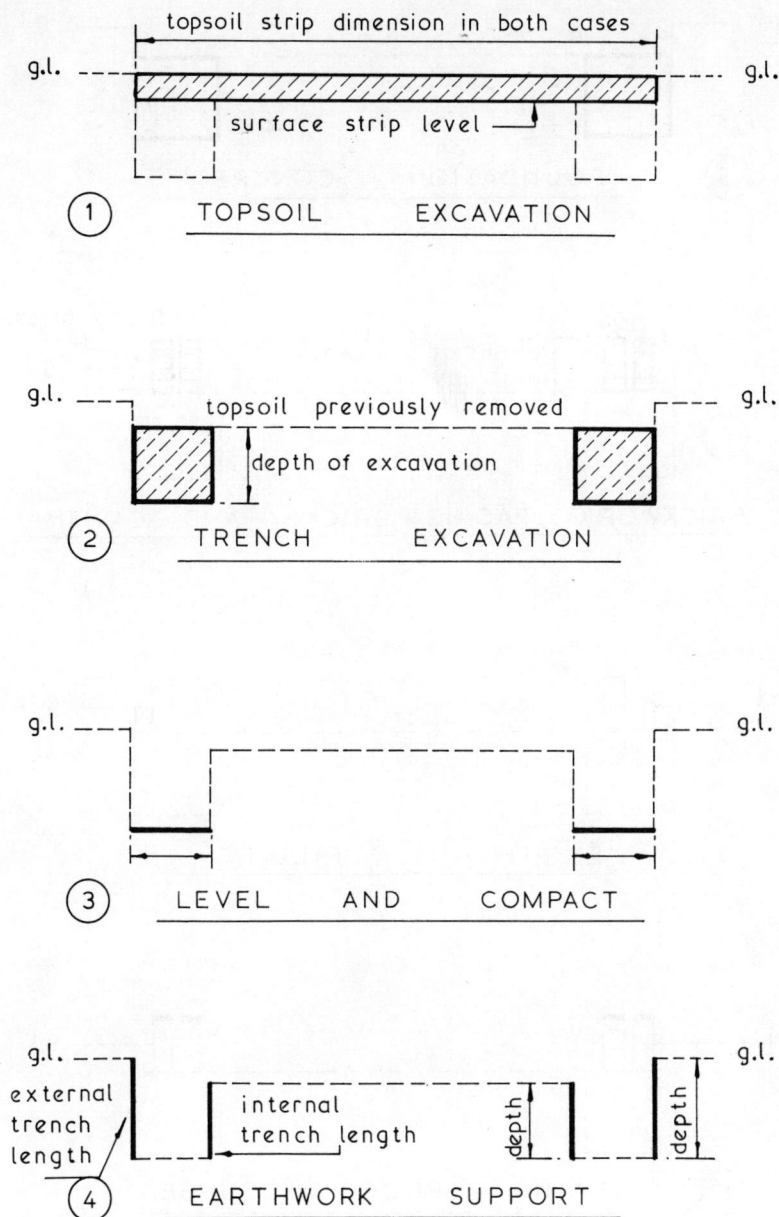

topsoil strip dimension in both cases

g.l.

surface strip level

1 TOPSOIL EXCAVATION

g.l. topsoil previously removed g.l.

depth of excavation

2 TRENCH EXCAVATION

g.l. g.l.

3 LEVEL AND COMPACT

g.l. g.l.

external trench length

internal trench length

depth

depth

4 EARTHWORK SUPPORT

Fig. 8.19 Example 8.10 Measurement sequence

142

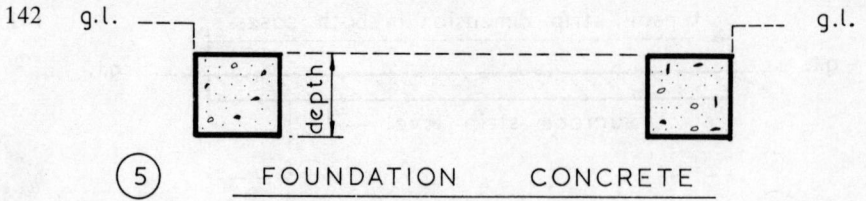

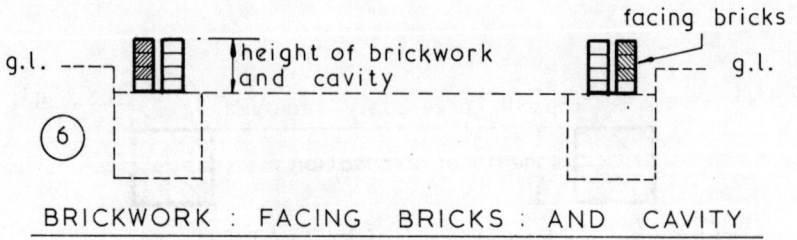

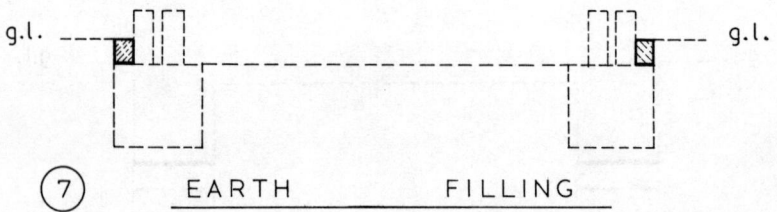

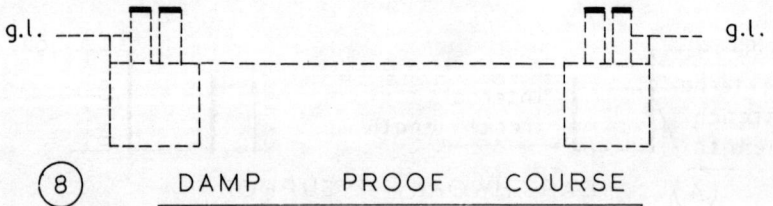

Fig. 8.20 Example 8.10: Measurement sequence

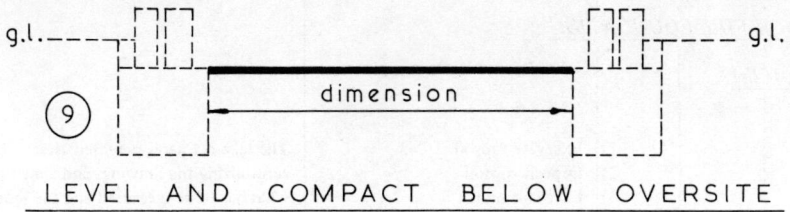

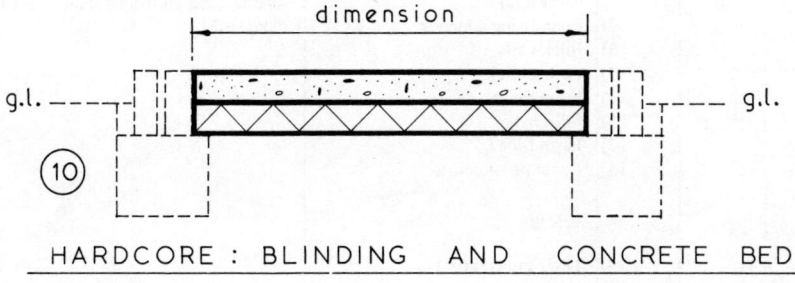

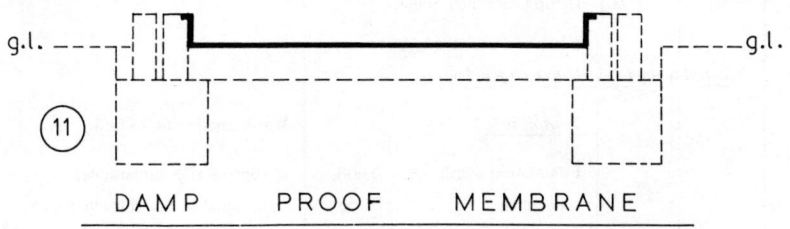

Fig. 8.21 Example 8.10: Measurement sequence

Take-off for Fig. 8.18

TRENCH FILL FOUNDATION

Take-off list

  1) Excavate topsoil
  2) Deposit topsoil
  3) Excavate trench
  4) Disposal of excavated material
  5) Earthwork support
  6) Level and compact base of
     excavation
  7) Foundation concrete
  8) Brickwork
  9) Facing bricks
10) Form cavity
11) Fill cavity
12) Earth filling
13) Damp proof course.

Oversite

14) Level and compact
15) Hardcore
    Blinding
    Concrete bed
    Tamp surface
    Fabric reinforcement
16) Damp proof membrane.

The take-off list is prepared after scrutinizing the drawings and aims to identify what has to be measured and the sequence of measurement.
A few minutes spent doing this will avoid many errors in practice.
As each item on the list is measured it can be ticked off.

The following in work up to d.p.c. level

Topsoil strip

Foundation width	=	0.600
wall width	=	0.255
projection	=	0.345

Waste calculation to find:

1. topsoil strip dimensions;
2. external trench side length;
3. centre-line of trench (this is the same as the cavity length in this case);
4. internal trench side length.

Strip length

Wall length	4.000
add projection	0.345
	4.345

Strip width

Wall length	2.500
add projection	0.345
	2.845

External trench length

2/4.345	8.690
2/2.845	5.690
	14.380

TRENCH FILL FOUNDATION continued

Centre line of trench

External trench length	14.380	Deduct from external trench length 2/300 per corner.
deduct 4/2/300	2.400	300 mm = half trench width.
	11.980	

Internal trench length

External trench length	14.380	Deduct from external trench length 2/600 per corner.
deduct 4/2/600	4.800	600 mm = trench width.
	9.580	

4.35 2.85	12.40	Excavate topsoil to be preserved average 150 mm deep.	**SMM D.9** Dimensions to extreme edge of foundation trench. Measured in m² stating average depth of excavation.
4.35 2.85 0.15	1.86	Deposit preserved topsoil in temporary spoil heap for re-use average 20 m from excavation.	**SMM D.31** The same dimensions are used as for topsoil excavation but depth added as required in m³.
11.98 0.60 0.75	5.39	Excavate foundation trench exceeding 0.30 m wide commencing at surface strip level maximum depth not exceeding 1.00 m	**SMM D.6.b, D.11** Centre line of trench × width of trench × depth of trench below surface strip level.
		&	
		Remove excavated material from site.	**SMM D.29**
14.38 0.90	12.94	Earthwork support maximum depth not exceeding 1.00 m and width between opposing faces not exceeding 2.00 m.	**SMM D.14, D.15, D.17** Maximum depth given as clause D.1 and width between opposing faces as D.17. The depth of support to external trench is 900 mm deep whilst the internal trench is 750 mm deep.
9.58 0.75	7.19		
	20.13		
11.98 0.60	7.19	Level and compact base of excavation to receive concrete.	**SMM D.40** Measured in m². Centre-line of trench × width of trench.
11.98 0.60 0.75	5.39	Plain concrete in foundation trenches (1 : 3 : 6 – 19) over 300 mm thick poured against face of excavation.	**SMM F.6.2, F.5.2, F.4.6** Measured in m³ stating thickness as clause F.5.2. Centre-line of trench × width of concrete × depth of concrete.

TRENCH FILL FOUNDATION continued

Brickwork in foundations

Centre line of external skin

Perimeter of		Waste calculation to find centre-line of
brickwork		external skin of brickwork.
2/4.000	8.000	
2/2.500	5.000	
	13.000	
deduct 4/2/51.25	0.410	
	12.950	

Centre line of internal skin

Waste calculation to find centre-line length of internal skin.

Note deduction =

perimeter of			external wall	= 102.5
brickwork	13.000		cavity	= 50
deduct			half internal skin	= 51.25
4/2/203.75	1.630			203.75
	11.370			

∴ deduct 2/203.75 per corner.

Two methods of measuring the brickwork are now described.
In this the first the centre lines of both the external and internal walls have been calculated. These have been timesed by the height of the common and facing bricks as appropriate.
Many building students find the first method easier to understand whilst quantity surveyors would probably use the second method.

11.37				*SMM G.5.1, G.5.3*
0.30	3.41	Half brick thick wall in skins of		Centre line of internal skin × height of brick-
		hollow wall in calcium silicate		work to d.p.c. level. Centre-line of external
12.59		bricks class 3 in stretcher bond in		skin × height of one course of brickwork.
0.08	1.01	cement mortar (1 : 4)		The other 3 courses of the external skin will
				be measured in facing bricks.
	4.42			

				SMM G.14.9, G.5.3
12.59		Half brick thick wall in skins of		Half brick and one brick thick walls entirely in
0.23	2.90	hollow walls entirely in facing		facing bricks are kept separate.
		bricks (P.C. £100 per 1000) in		Note description in skins of hollow walls.
		stretcher bond in cement mortar		Centre-line of external skin × height of 3
		(1 : 4) pointed with a neat rubbed		courses of facing bricks. Height of one course
		in joint.		of brickwork = 65 mm + 10 mm joint
				= 75 mm

			ALTERNATIVE METHOD	
2/	11.98 0.30	7.19	Half brick thick wall in skins of hollow walls in calcium silicate bricks class 3 in stretcher bond in cement mortar (1 : 4)	Here the centre-line of the cavity has been twice timesed. This gives the actual length of both skins. This can only be done where both skins have the same width dimensions.
	12.59 0.23	2.90	*Deduct* ditto & *Add* Half brick thick wall in skins of hollow walls entirely in facing bricks (P.C. £100 per 1000) in stretcher bond in cement mortar pointed with a neat rubbed in joint.	This adjustment is necessary to obtain the correct quantity of common and facing bricks. In the previous dimension the total brickwork was measured to d.p.c. level as if all in sand-limes. The adjustment corrects the quantities for commons and measures the area of facing bricks.
	11.98 0.30	3.59	Form 50 mm wide cavity in hollow wall including 4 number galvanized butterfly wall ties per m².	*SMM G.9* Given in m² stating width of cavity and type and spacing of wall ties in the description. Centre-line of cavity × height of cavity to d.p.c. level.
	11.98 0.05 0.15	0.09	Weak concrete (1 : 12) filling to hollow wall not exceeding 100 mm thick.	*SMM F.6.1, F.6.18, F.5.1* Concrete filling to cavity measured in m³ stating width as clause F.5.1. Cavity fill taken in this case to within 150 mm of d.p.c. level.
			External trench length = 14.380 Face of brickwork perimeter = 13.000 2)27.380 13.690	Waste calculation to find centre-line length of earth filling. External trench length and face brickwork perimeter are added together. The centre-line of earth fill is the mean of these two lengths.
			Width of earth fill foundation width = 600 wall width = 255 2) 345 projection = 172.5	Calculation to find width of earth filling.
	13.69 0.17 0.15	0.35	Earth filling around excavations with topsoil from temporary spoil heaps.	*SMM D.33, D.34* Measured in m³. Source of filling material to be classified as D.33.

TRENCH FILL FOUNDATION continued

Oversite construction

		Strip length	4.350
		deduct 2/600	1.200
			3.150
		Strip length	2.850
		deduct 2/600	1.200
			1.650

Waste calculation to find dimensions of level and compact under hardcore bed. Trench width is deducted from strip length × 2 for each length.

3.15		Level and compact base of
1.65	5.20	excavation to receive hardcore.

SMM D.40
Measured in m².

Length of
building   = 4.000
deduct
walls 2/255   = 0.510
             3.490

Width of
building   = 2.500
deduct
walls 2/255   = 0.510
             1.990

Waste calculation to find internal sizes of building.

Many of the oversite measurements are similar and can be anded on in the following manner.

3.49		Hardcore filling in making up
1.99	6.95	levels average 150 mm thick
		obtained off site

&

Blind surface of hardcore with sand

&

Reinforced concrete bed
(1 : 2 : 4 – 19) 100–150 mm
thick laid on blinded hardcore
bed (measured separately)
(Cube × 0.15 = 1.04 m³)

&

Tamp surface of unset concrete

&

Fabric reinforcement to
B.S. 4483 weighing 3.05 kg/m²
with minimum 150 mm end and
side laps in ground slab (measured
net no allowance for laps)

SMM D.33, D.36
Over 250 mm thick measured in m³. Under 250 mm thick measured in m² stating average thickness, and source of supply.

SMM D.43
Measured in m². Alternatively this could be included with filling where measured in m².

SMM F.4.2, F.4.6, F.6.1, F.6.8, F.5.2
Note concrete bed measured in m³. For convenience it has been anded on amongst other superficial dimensions. When squaring is carried out the cubic quantity is entered in the brackets.

SMM F.9
Treating surface of unset concrete given in m² stating type of treatment.

SMM F.12.1, F.12.2, F.12.3, F.11.4
Measured in m². State weight and side and end laps. No allowance for cover to reinforcement has been made here.

		Damp-proof membrane		Waste calculation to find sizes of damp-proof
		Internal length		membrane. 50 mm has been added to the internal
		of building	3.490	dimension of building to allow d.p.m. to turn
				under d.p.c.
		turn in under		
		d.p.c. 2/50	0.100	
			3.590	
		Internal width		
		of building	1.990	
		turn in under		
		d.p.c. 2/50	0.100	
			2.090	
		Vertical girth		
		Brick perimeter	13.000	Waste calculation to find length of d.p.m. placed
				vertically.
		Deduct corners		
		4/2/255	2.040	
			10.960	

SMM G.37.1, G.37.2

3.59	1000 gauge polythene	There is no specific mention in the SMM of damp-
2.09	damp-proof membrane laid	proof membranes. This description has been
	horizontally on blinded hard-	phrased using G.37 damp-proof courses.
	core bed with minimum	Note the quantity is measured net, the estimator
	200 mm laps (measured net	must add for laps.
	no allowance for laps)	

SMM G.37.1, G.37.2

10.96	Ditto but vertically to face	The vertical part of the d.p.m. does not exceed
	of brickwork 150 mm deep.	225 mm wide and is measured in linear metres.

Example 8.11
Timberwork generally

Having spent a considerable time on foundation work and structural walls the next logical step is to consider timber suspended floors and flat roofs which introduces the student to section N of the SMM entitled 'Woodwork'.

Clause N.1.1 of the SMM requires that the following general information be given:

(a) kind and quality of material including whether sawn or wrought;
(b) any preliminary treatment of material required;
(c) selection and protection for subsequent treatment;
(d) matching of grain or colour;
(e) surface treatment to be applied as part of the production process.

Sawn and wrought timber
The SMM requires that it should be stated if timber is to be sawn or wrought as there is a considerable difference in price.

Sawn timber
Sawn timber has a rough surface left by the saw at the mill and is normally used in carcassing work such as floor joists, partitions and roof members. The timber section in Fig. 8.22 would be described as 50 × 100 mm sawn.

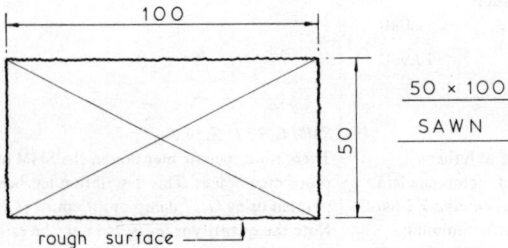

Fig. 8.22 Sawn timber

Wrought timber
Wrought timber has a smooth surface finish achieved by passing sawn timber through a planing machine at the mill. During this process the rough surface is removed on all four edges, so reducing the section size of the timber. Thus wrought timber is commonly described as being of *nominal* size. In practice this means that the timber section shown in Fig. 8.23 if described as

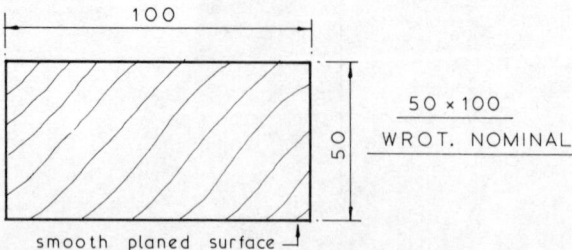

Fig. 8.23 Wrought timber

50 ×100 mm wrot *nominal* size would in fact be slightly smaller after being
machined. (Wrot is the abbreviated term for wrought.)

Should the timber be required as actual size of 50 × 100 then it should be described as **finished** size and would be machined down from a larger section of timber. Note that timber merchants often use the term 'prepared' which also means wrought.

Softwood and hardwood

It should always be stated whether the timber required is to be softwood or hardwood. Hardwood is considerably more expensive than softwood.

Jointing

SMM N.1.3 states that where no method of jointing is described then it is at the contractor's discretion.

Fixing

Fixing of timber is deemed to be by nails unless otherwise described SMM N.1.4.

Timber lengths

The quantity surveyor will measure the length of timber as fixed in the work. An allowance should be made for any specified joint such as a tenon or a mortise and tenon joint; a halved joint on wall plates or a tusk tenon joint.

The fact that the timber is measured the actual length in the work is very important to the contractor because he will purchase in stock lengths. For example although a floor joist may be 3.10 m long in the work the next stock size above is 3.30 m and that will be the length that the contractor must pay for. The estimator must allow a wastage factor when pricing such items. Timber stock sizes are:

1.8 m	3.3 m	4.8 m
2.1 m	3.6 m	5.1 m
2.4 m	3.9 m	5.4 m
2.7 m	4.2 m	5.7 m
3.0 m	4.5 m	6.0 m
		6.3 m

The SMM clause N.1.6 requires that timber required to be in one continuous length exceeding 4.20 m for softwood and 3.00 m for hardwood shall be given in further stages of 300 mm. The reasons for this is that larger lengths of timber cost extra and are more difficult in terms of site handling. This clause is illustrated in Fig. 8.24.

152

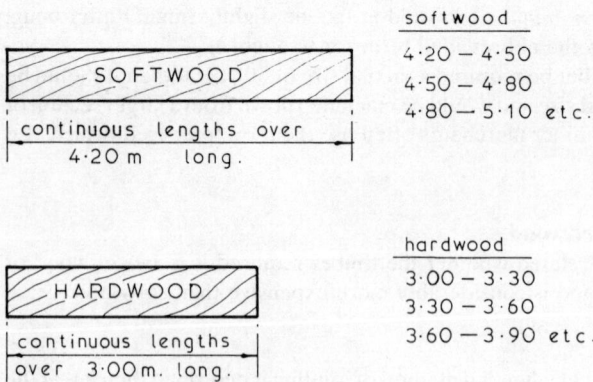

softwood

4·20 – 4·50
4·50 – 4·80
4·80 – 5·10 etc.

SOFTWOOD

continuous lengths over
4·20 m long.

hardwood

3·00 – 3·30
3·30 – 3·60
3·60 – 3·90 etc.

HARDWOOD

continuous lengths
over 3·00m. long.

Fig. 8.24 Timber length

Cutting and forming ends angles and mitres

Clause N.1.10 states that labour items such as fitting ends, cutting angles and mitres by the carpenter are not measured unless the sectional area of the timber member involved exceeds 0.002 m² in which case they shall be enumerated. Figure 8.25 illustrates this point.

The student should now work through the timber suspended floor and roof examples that follow.

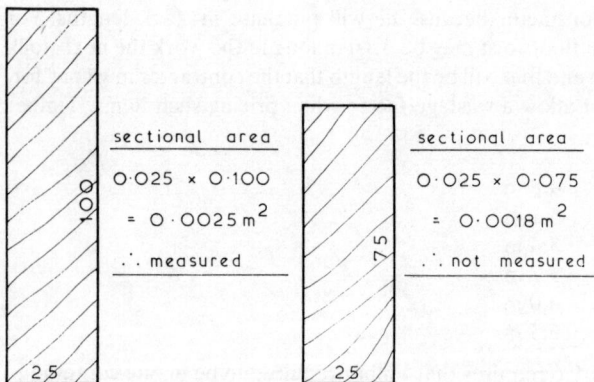

sectional area

0·025 × 0·100

= 0·0025 m²

∴ measured

sectional area

0·025 × 0·075

= 0·0018 m²

∴ not measured

Fig. 8.25 Ends, angles and mitres

Example 8.12.
Timber suspended floor

The example given is for a simple construction that will be found in housing projects or similar situations. In measurement terms the floor may be broken into two headings namely:

1. floor construction;
2. floor coverings.

The main problem is to calculate the number of floor joists and their lengths. On more involved situations a combination of calculation and scaling from drawings may be appropriate, particularly where there are double joists and trimming for obstructions encountered and these will be considered at Level III.

Reference should now be made to Figs. 8.26 and 8.27 and to the specification details that follow.

Specification details

Floor joists   50 × 200 mm sawn softwood at 400 mm centres.

Strutting   38 × 38 mm sawn softwood herring-bone strutting.

Floor covering   18 mm thick tongued and grooved flooring grade chipboard.

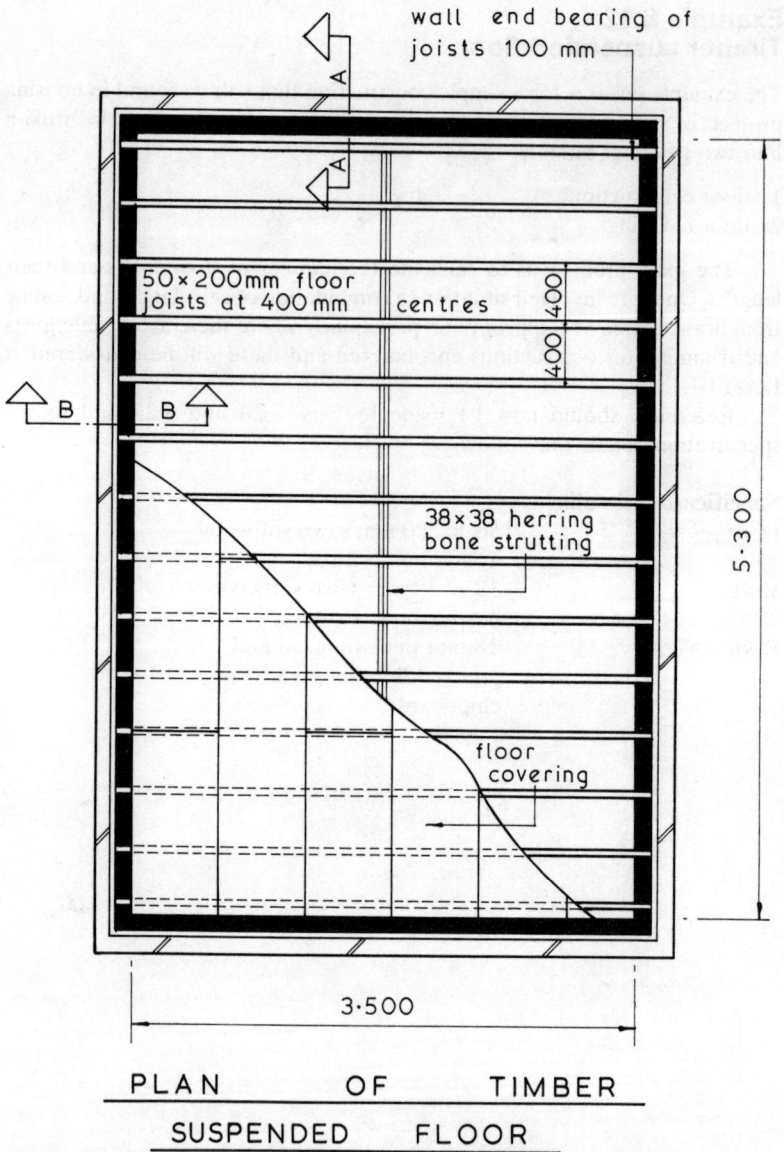

Fig. 8.26 Example 8.12: Timber suspended floor

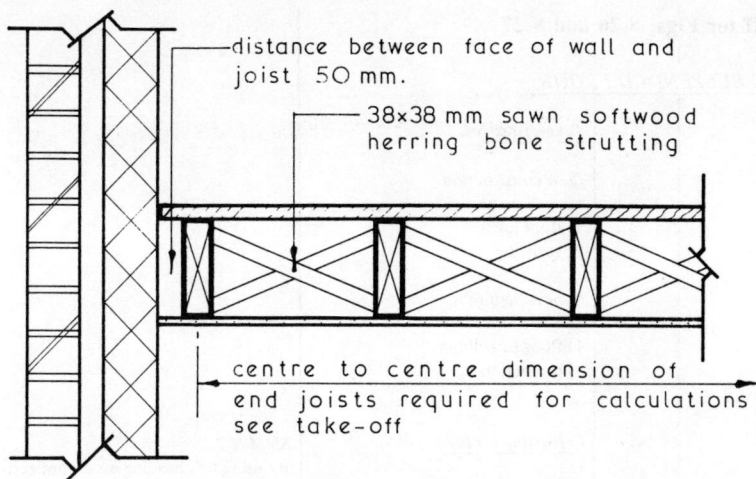

section A-A suspended timber floor.

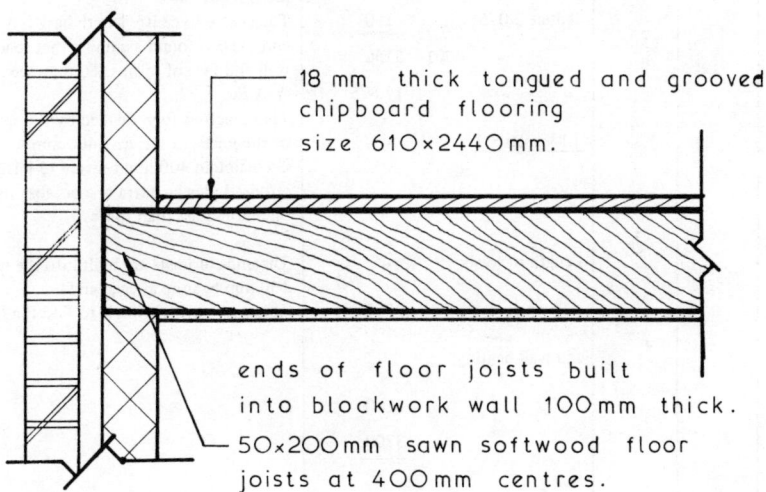

section B-B suspended timber floor.

Fig. 8.27 Example 8.12: Details of timber suspended floor

Take-off for Figs. 8.26 and 8.27

TIMBER SUSPENDED FLOOR

Take-off list

floor construction

1) floor joists
2) strutting

floor coverings

1) floor boarding
2) protection

Carcassing timber

SMM N.2
By using this heading we do not constantly have to state in descriptions that these are carcassing items.

			length of room		5300
			deduct 2/075		150
				400)	5150
			=		12.875
			∴ 13 joists + 1		

Waste calculation to find number of floor joists required.
The centre to centre length between the two end joists is found assuming that space between wall and face of joists is 50 mm, see section A–A Fig. 8.27.
That length is then divided by the spacing of of the joists, in this case 400 mm.
To maintain 400 mm centres 13 joists are required plus an extra one because there will be one more joist than spaces.

Length of joists

clear span   3500

add wall bearing

2/100   200

     3700

The ends of joists are built into the wall with a 100 mm bearing at each end.
This waste calculation is to find the total length

14/	3.70	51.80	50 × 200 mm sawn softwood in floors.

SMM N.2.1
Carcassing timber measured in linear metres stating cross sectional size.
Classified as in:
(a) floors;
(b) partitions;
(c) flat roofs;
(d) pitched roofs including ceiling joists;
(e) kerbs bearers and the like.

	5.30	5.30	38 × 38 mm sawn softwood herring bone strutting to 200 mm deep joists.

SMM N.2.2
Measured in linear metres on the horizontal plan length.
Depth of joists to be stated.

TIMBERED SUSPENDED FLOOR continued

			Floor coverings	
			First fixings	SMM heading used to save necessity of repeating this in description.
	5.30 3.50	18.55	18 mm thick tongued and grooved flooring grade chipboard in floors fixed to joists with 65 mm annular ringed shank nails	*SMM N.4.1* Measured in m². Thickness and method of jointing and fixing to be given in description. Classified as N.4.1 which in this case is floors.
			& Allow for protection to ditto.	*SMM N.33.1* Protection to the area of flooring through the course of the work shall be given as an item stating the area. The use of protection items such as these is fully described in Example 8.14.
			Alternative description for board flooring.	
			18 mm tongued and grooved softwood wrought board flooring well cramped up and fixed to each joist with two number 65 mm floor brads	*SMM N.4.1* Alternative description where tongued and grooved board flooring is specified. The take-off is now billed and the student should note how again the SMM is used to determine the order and headings of the bill.

Bill of quantities for Figs. 8.26 and 8.27

WOODWORK

Carcassing

Sawn softwood

A	50 X 200 mm in floors.	52	m
B	38 X 38 mm herring bone strutting to 200 mm joists.	5	m

First fixings

Boardings and flooring

C	18 mm tongued and grooved flooring grade chipboard in floors fixed to joists with 65 mm annular ringed shank nails	19	m²

Protection

D	Allow for protection to chipboard flooring area 19 m²	item	
		To summary	£

Note: To summary at end of section of bill, otherwise to collection, see Example 8.14.

Example 8.13.
Timber flat roof

The example given is for a timber flat roof construction to a small building. The roof joists are laid level and the fall of the roof is obtained by the firring pieces.

It would be the normal practice after measuring the construction of the roof to deal with the coverings which would most likely be built up felt in this case. This is, however, felt to be outside the basic level of the syllabus. For similar reasons rainwater goods are not measured.

The opportunity is taken to introduce the student to the painting and decorating section of the SMM in measuring the painted finish to the fascia board.

Figure 8.28 should now be studied together with the specification details that follow.

Specification details

Roof joists	50 × 125 mm sawn softwood at 400 mm centres.
Firring pieces	50 mm sawn softwood fixed to top of roof joists tapering 50 mm to nil.
Roof decking	18 mm thick roofing grade butt jointed chipboard.
Tilting fillet	38 × 38 mm sawn softwood triangular tilting fillet.
Drip	25 × 50 mm sawn softwood drip.
Fascia board	25 × 225 mm wrought softwood to three sides of roof. 25 × 150 mm ditto to rainwater discharge side.
Painting to fascia board	Knot, prime, stop, two undercoat and one gloss to front and bottom face of fascia board. Prime only back and top face of fascia prior to fixing.

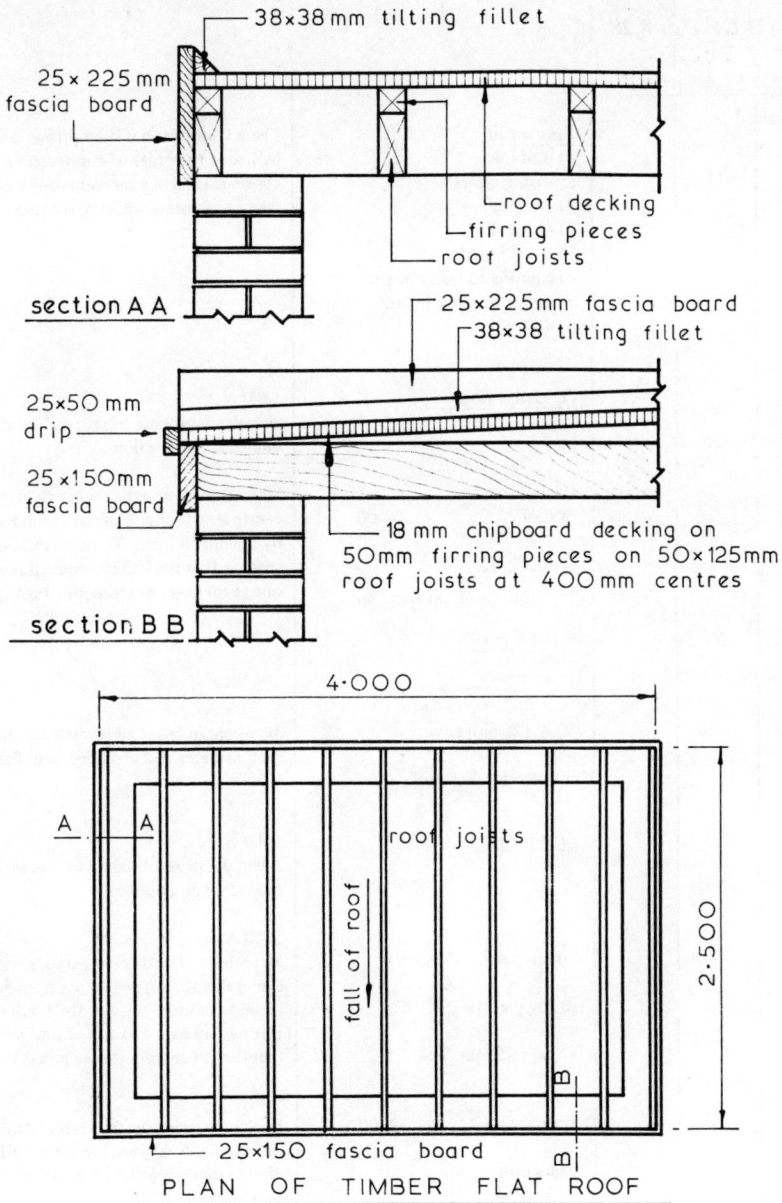

Fig. 8.28 Example 8.13: Timber flat roof

Take-off for Fig. 8.28

TIMBER FLAT ROOF

Take-off list
1) roof joists
2) firring pieces
3) decking
4) tilting fillet
5) fascia board
6) painting to fascia board
7) mitres to fascia board
8) drip

The take-off list has been prepared in the m
by listing the order of construction. It is
convenient from a measurement viewpoint
take for painting whilst measuring the fascia
board.

Carcassing timber

SMM N.2
Use of sub-heading to save inclusion in
individual descriptions.

Overall length
of roof   4000

deduct 2/025   50

         400) 3950

= 9.875 = 10 + 1

Similar calculation to timber floor. Centre t
centre length of end joists is found and divic
by spacing of joists. To maintain a 400 mm
spacing 10 joists will be required plus 1 extr
one as there is one more joist than spaces.

11/	2.50	50 × 125 mm sawn softwood in flat roof.

SMM N.2.1.c
Measured in linear metres stating cross-
sectional area and classified as in flat roof.

First fixings

SMM N.4
Again use of sub-heading to save inclusion in
individual descriptions.

11/	2.50	50 mm sawn softwood firring pieces average 25 mm deep.

SMM N.6.1
Measured in linear metres stating width and
average depth. Firrings taper from 50 mm to
nil and therefore the average depth is 25 mm
Firrings are nailed to top of roof joists and
their length and number required are the sam

Width of roof   2500

add width of

fascia   25

           2525

Decking passes over the fascia board on
discharge side of roof, see section BB.
Waste calculation to find actual width of
decking.

4.00		18 mm thick butt jointed roofing grade chipboard fixed to softwood bearers in flat roof.
2.53		

SMM N.4.1, N.4.1.d
Measured in m² stating thickness and method
of jointing and fixing. Classified as N.4.1.d in
flat roofs.
Roofing grade chipboard can be obtained wit
a single layer of felt already applied and this
should be specified where required.

TIMBER FLAT ROOF continued

2/	2.53 4.00	38 × 38 mm sawn softwood tilting fillet.	*SMM N.10.1* Measured in linear metres stating cross-sectional area. Note that the fillet is required to three sides of the roof but not to the rainwater discharge side.

Length of

brickwork   4000

add mitres

2/025   50

                4050

Width of

brickwork   2500

add mitres

2/025   50

                2250

Waste calculation to find actual length of fascia board to allow for cutting of mitres at the corner of the roof.

2/	4.05 2.25	25 × 225 mm wrought softwood fascia board in one width	*SMM N.4.1.g, N.1.5* Measured in linear metres stating cross-sectional dimensions. N.1.5 requires that except for carcassing items, members over 200mm wide required to be in one width shall be so described.

                            &

Knot, prime, stop

② & ① to ditto

150-300 mm girth new work

externally

                            &

Prime only back of fascia prior

to fixing 150-300 mm girth

new work externally

SMM V.3.1, V.3.2
Painting to isolated surfaces not exceeding 300 mm girth given in linear metres stating girth range as
0-150 or 150-300.
2 means 2 undercoats
1 means 1 gloss.
Note classification of work as clause V.3.1.

Girth for painting =
front face   225
bottom edge   25

                  250

(this is girth range 150-300)
Girth for priming =
back face   225
top edge   25

                250

(also girth range 150-300)

TIMBER FLAT ROOF continued

4.05			25 × 150 mm wrought softwood fascia board.	*SMM N.4.1.g* Similar to before but smaller fascia to rain-water discharge side. Clause N.1.5 does not apply in this case as width does not exceed 200 mm.

&

Knot, prime, stop

② and ① to ditto

150-300 mm girth new work

externally.

SMM V.3.1, V.3.2
Girth

Front face of fascia	150
Bottom edge of fascia	25
	175

∴ girth range 150-300.

&
Prime only back of fascia prior

to fixing 150-300 mm girth new

work externally.

Girth

Front face of fascia	150
Bottom edge of fascia	25
	175

∴ girth range 150-300.

4		Mitres to fascia board.

SMM N.1.10
Mitres on timber are deemed included except where cross-sectional area of member exceeds 0.002 m² in which case they are enumerated.
0.025 × 0.225 = 0.005 m²
0.025 × 0.150 = 0.003 m²
Both exceed 0.002 m² sectional area and are measured.

4.05		25 × 50 mm sawn softwood drip.

SMM N.6.2
Measured in linear metres stating cross-section dimensions.

This example is not billed and reference shou be made to Example 8.14 which is similar.

Example 8.14.
Model take-off and bill for complete structure

The purpose of this example is to group together into one structure much of the measured work that has been studied in the previous examples. We can then build upon this knowledge during Level III so that at the end of that period the student should have enough confidence and ability to measure quantities for simply constructed buildings.

The taking-off in this example is handwritten using abbreviated descriptions because this is what the student must do in practice. A list of suggested abbreviations is published by the Institute of Quantity Surveyors, and students who intend pursuing a career related to this subject could usefully adopt these. Abbreviations are used to reduce the time spent in writing descriptions during taking off, but are not used in the bill of quantities in case of misinterpretation.

The SMM requires that certain details be given in the form of an item in the bill of quantities. An item will be priced on a lump sum basis after the estimator has evaluated the rest of that section of the bill.

Typical examples of items are for bringing to and removing from site plant required for a particular section of the work and for its maintenance whilst on site. At the end of each work section a protection item is given to allow the estimator to price for any protective coverings or measures that may be necessary for that part of the work.

Reference should now be made to the section and plan of the building shown in Figs. 8.29, 8.30 and 8.31 and to the specification details that follow. By now students will have realized that a good knowledge of building technology is essential in measurement as often drawings do not convey the level of detail that is required, and therefore the taker-off has to rely on his own judgement.

Finally, after the take-off a bill of quantities has been prepared. This has been kept as simple as possible to encourage the student to undertake similar exercises, including where possible elementary pricing. Although not done in this example the student should be aware that excavation, concrete and brickwork in foundations is often grouped together in one complete bill section as sub-structure work.

Note also that the dimension for weak concrete filling to hollow wall when squared up equals 0.14 m³. When billed this quantity is rounded to the nearest metre and would therefore be eliminated. Clause SMM A.7.3 makes provision for this by stating that in such cases the item should be enumerated stating the size.

Specification details
Sub-structure

Excavation	Topsoil to be retained on site for future use. Trench excavation to be removed from site.
Foundation concrete	Plain mass concrete (1:3:6 – 19)
Oversite construction	Plain concrete bed (1:2:4 – 19) 150 mm thick incorporating 1000 gauge polythene damp-proof membrane laid on 150 mm thick hardcore bed blinded with sand.

Cavity filling

Brickwork

Damp proof course

Super-structure
Brickwork

Timber flat roof

Weak concrete fill to cavity (1:12)
taken to within 150 mm of damp-proof
course level with a splayed top edge.
Foundation brickwork to be 255 mm wide
cavity construction, consisting of
two 102.5 mm wide skins of brickwork in
calcium silicate bricks class 3
in cement mortar (1:4). Cavity 50 mm
wide including 4 number galvanized
butterfly wall ties per m². For type of
facing bricks see super-structure
brickwork.
102.5 mm wide hessian-based bituminous
felt d.p.c. to BS 743, Part F in single
layer with minimum 100 mm laps bedded
in cement mortar (1:4).

Cavity construction. External skin of
102.5 mm Wealden stock facing bricks in
stretcher bond in gauged mortar (1:1:6)
pointed with a neat rubbed-in joint
extending to three courses below d.p.c.
level.
Internal skin of 100 mm thick solid
Celcon blockwork keyed for plaster
and bedded in gauged mortar (1:1:6).
Cavity 50 mm wide including 4 number
galvanized butterfly wall ties per m².

50 × 225 mm sawn softwood roof joists
at 400 mm centres.
Firring pieces 50 mm sawn softwood
tapering 75 mm to 0.
18 mm thick plain edged roofing grade
chipboard roof decking.
38 × 38 mm sawn softwood triangular
tilting fillet.
Fascia boards 18 mm thick WBP (weather
and bailproof) plywood. 385 mm deep to three
sides of roof and 250 mm deep
to rainwater discharge side. Painted
finish to fascia consisting of knot,
prime, stop and two oils. Back of
fascia to be primed prior to fixing.

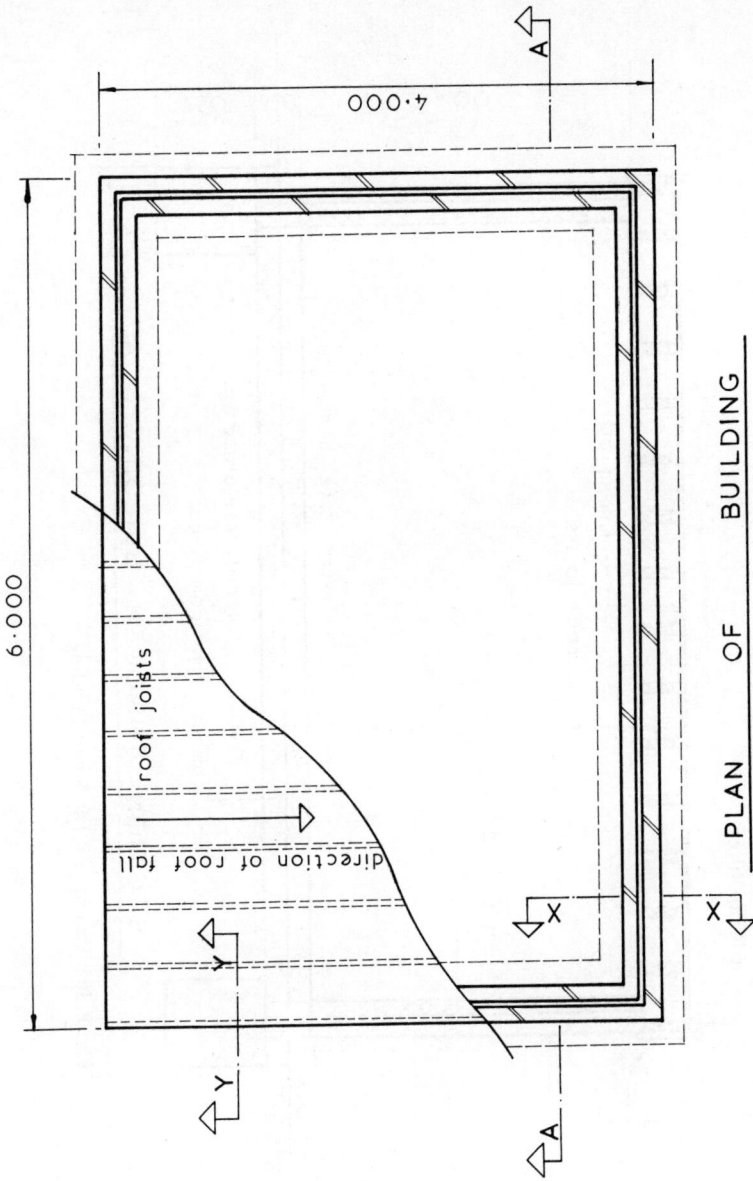

Fig. 8.29 Example 8.14: Small building

166

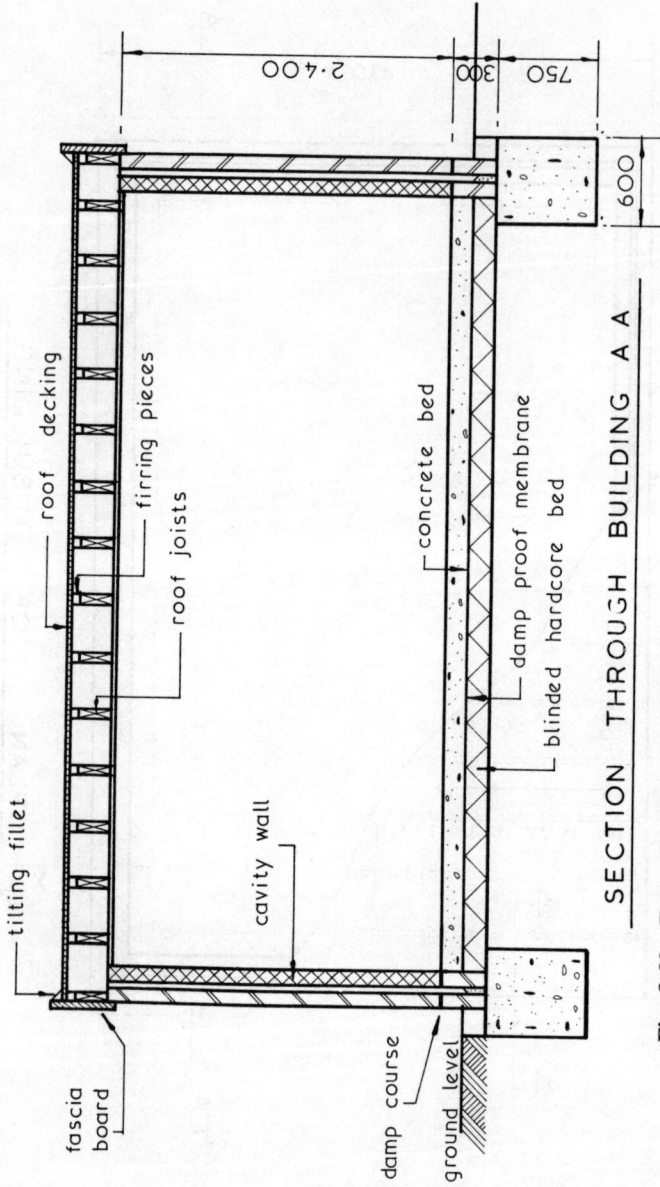

Fig. 8.30 Example 8.14: Small building

Labels on figure:
- 2.400
- 300
- 750
- 600
- roof decking
- firring pieces
- roof joists
- concrete bed
- damp proof membrane
- blinded hardcore bed
- tilting fillet
- cavity wall
- fascia board
- damp course
- ground level

SECTION THROUGH BUILDING A A

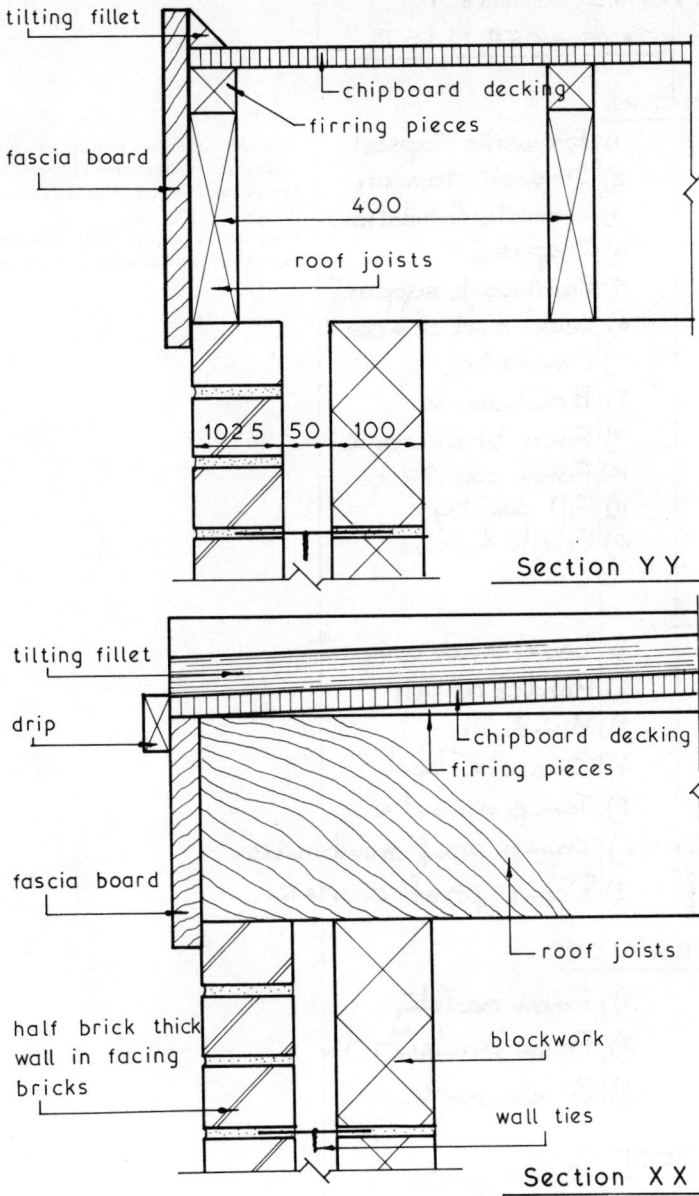

Fig. 8.31 Example 8.14: Eaves details

Take-off for Figs. 8.29, 8.30 and 8.31

TAKE - OFF LIST

SUB-STRUCTURE

1) Excavate topsoil
2) Deposit topsoil
3) Excavate foundation
4) Disposal
5) Earthwork support
6) Level and compact
7) Concrete
8) Brickwork
9) Face brickwork
10) Form cavity
11) Fill cavity
12) Earth filling

OVERSITE

1) Level and compact
2) Hardcore fill
3) Blinding
4) Concrete bed
5) Tamp concrete
6) Damp proof membrane
7) Damp proof courses

SUPER-STRUCTURE

1) Form cavity
2) Face brickwork
3) Blockwork

FLAT ROOF

Roof joists
Firring pieces
Roof decking
Tilting fillet
Fascia board
and painting
Mitres to fascia bd.
Drip

/.

A take-off list has been prepared here for the whole structure. Alternatively, it could have been prepared for each individual section of the work.
Remember the purpose is to think out on paper what has to be subsequently measured.

Substructure

__Item__	Allow for bringing to and removing from site all plant req'd for excavation and earthwork	

SMM D.4, F.2, G.3
Plant items are required in this example for excavation and earthwork, concrete work and brickwork and blockwork.

&

Do. concrete wk.

&

Do. brick. & block

__Item__	Allow for maintaining on site all plant req'd. for excav'n. and earthwork.	

&

Do. concrete wk.

&

Do. brick & block.

Topsoil Strip

Fdn . width	600
Wall . width	255
Proj .	345

Waste calculation to find topsoil strip dimensions.

2.

Strip length

Blg. lgth.	6.000
Add proj.	0.345
	6.345

Strip width

Bldg. lgth.	4.000
Add proj.	0.345
	4.345

6.35		Excavate topsoil
4.35	27.62	to be preserved average 150 mm dp.

SMM D.9
Measured in m² stating average depth.

6.35		Deposit pres.
4.35		topsoil in temporary
0.15	4.14	spoil heap for re-use average 25 m from excav'n.

SMM D.31
Measured in m³ stating average distance of spoil heap in metres or kilometres from excavation point. Alternatively, this could have been anded on to previous description and provision made for cubing dimensions.

Trench Excavation

Centre line trench

2/6.345	12.690
2/4.345	8.690
	21.380
Adj. corners	
Ddt. 4/2/300	2.400
	18.980

Waste calculation to find perimeter length of external trench.
Corners are then deducted to find centre-line length of trench in the normal way. For example half trench width (300) × 2 per corner and four corners.

3.

18.98		Excavate found/. trench exceeding 0.30 m wide commencing at surface strip level maximum depth not exceeding 1.00m
0.60		
0.75	8.54	

SMM D.10, D.11, D.6
Trenches over 0.30 m wide measured in m^3.
State starting level and maximum depth of
trench as D.11.
Max. depth not exceeding 0.25 m
Max. depth not exceeding 1.00 m
Max. depth not exceeding 2.00 m
Max. depth not exceeding 4.00 m
and thereafter in 2.00 m stages.

SMM D.29
Disposal also in m^3 and can be anded on.

&

Remove excavated material from site

Earthwork Support

Internal trench lgth.

Ext. tr. lgth.	21.380
Ddt. corners.	
4/2/600	4.800
	16.580

Waste calculation to find internal trench
length. The external length was previously
calculated as 21.380. From this corners can
be deducted. In this case the whole trench
width 600 X 2 per corner and four corners.

Trench depth

Topsoil	150
Trench dig	750
	900

21.38		Earthwork support max. depth n.e. 1.00 m and width between opposing faces n.e. 2.00 m.
0.90	19.24	
16.58		
0.75	12.44	
	31.68	

SMM D.14, D.15, D.16, D.17
Measured in m^2 stating maximum depth as
D.11 and width as D.17. The dimensions
used are external trench perimeter X vertical
height and the internal trench perimeter X
vertical height.

4.

18.98			Level and compact
0.60	11.39		base of excavation
			to recieve concrete

SMM D.40
Measured in m² . Centre-line of trench ×
width of trench.

Foundation Concrete

18.98		Plain conc. in
0.60		foundations (1:3:6-19)
0.75	8.59	over 300 mm
		thick poured ag.
		face of excau'n .

SMM F.6.2, F.5.2, F.4.2
Measured in m³ stating thickness as clause
F.5.2. Centre-line of trench length × width
of concrete × depth of concrete.

Brickwork in foundations .

Centre line length
external skin .

2/6.000	12.000
2/4.000	8.000
Ext. per.	20.000
Ddt. corners	
4/2/51.25	0.410
	15.590

Waste calculation to find centre-line of
external skin of brickwork. Half of external
skin width 51.25
∴ deduction per corner = 2/51.25 .

C.L. of cavity

Ext. per.	20.000
Ddt. corners	
4/2/127.5	1.020
	18.980

The centre line of cavity should be the same
as the centre-line of the trench and this
calculation confirms that fact.Build up:

Width of external skin	102.5
Width of half cavity	25
Deduction per corner	2/127.5

5.

C.I. internal skin.

Ext. per.   20.000
Ddt. corners
4/2/203.75   1.630

               18.370

Waste calculation to find centre-line length
of internal skin.
Build up:

External skin width	102.5
Cavity	50
Half internal skin width	51.25

Deduction per corner   2/203.75

Brickwork height

To d.p.c.   300
Facings   225

             075

Waste calculation to find height of facing
bricks below d.p.c. level.

19.59
0.08 1.57   Half B. th. wall
             in skins of hollow
(18.3)   walls in calcium
0.30 5.51   silicate class 3
             bricks in stretcher
             bond in c.m. (1:4)

7.08

SMM G.5.1, G.5.3
Measured in m² length of external skin
× height of one course of brickwork.
Length of internal skin × height of four
courses of brickwork.

Brick facework

19.59
0.23 4.50   Half. B. th. wall
             in S.O.H.W.
             entirely in facing
             bricks Wealden
             stocks in streth.
             bond in c.m. (1:4)
             pointed wi. a
             rubbed in joint.
             6.

SMM G.14.9, G.5.3
Half brick and one brick thick walls built
entirely in facing bricks are kept separate.
Note description in skins of hollow walls.
Centre-line of external skin × height of
3 courses of facing bricks.

Cavity

18·98		
0·30	5·69	Form 50 mm wide cavity in hollow wall incl'd. 4 no. galvanized butterfly wall ties to B.S. 1243 per m².

SMM G.9
Given in m² stating width of cavity and type and spacing of wall ties in the description. Centre-line of cavity × height to d.p.c. level.

18·98		
0·05		
0·15	0·14	Weak concrete filling (1:12) to hollow wall not exceeding 100 mm thick.

SMM F.6.1, F.6.18, F.5.1
Concrete filling to cavity measured in m³ stating width as clause F.5.1. Cavity fill taken in this case to within 150 mm of d.p.c. level. Squared dimension under 1 m³ and therefore enumerated on bill. See SMM clause A.7.3.

Earth Filling

Ext. per. tr.   21·380
Ext. per. bwk. 20·000

2) 41·380

20·690

Waste calculation to find centre-line length of earth filling. External trench length and brickwork external length perimeter added together. The centre-line of earth fill is the mean of these two lengths.

Fdn. width   600
Wall width   255

2) 345

172·5

Waste calculation to find width of earth filling.

20·69		
0·17		
0·15	0·53	Earth filling around excavations with topsoil from temporary spoil heap.

SMM D.33, D.34
Measured in m³. Source of filling material to be classified as D.33.

7.

Oversite construction.

			Level & compact.	Waste calculation to find sizes of dimensions for area of level and compact.
			Trench lgth. 6·345	
			Ddt.	
			2/600 1·200	
			___5·145___	

			Trench lgth. 4·345	Trench width (600) is deducted 2 × from strip length for each dimension.
			Ddt.	
			2/600 1·200	
			___3·145___	

5·15			Level and	*SMM D.40*
3·15	16·22		compact ~~base~~	Measured in m². Length × width of surface below hardcore fill.
___			of excavation	
			to receive	
			hardcore filling	

			Bldg. lgth. 6·000	Waste calculation to find dimensions of oversite construction. The internal dimensions of building between external walls are required.
			Ddt. walls.	
			2/255 0·510	
			___5·490___	

			Bldg. widlk 4·000
			Ddt. walls
			2/255 0·510
			___3·490___

8 .

5.49		
3.49	19.16	

Hardcore filling obtained off site in making up levels average 150 mm thick.

SMM D.33, D.36
Measured in m² stating average thickness. Over 250 mm thick measured in m³. Source of supply to be classified as D.33.

&

Blind surface of hardcore with sand

SMM D.43
Measured in m².

&

Plain concrete bed (1:2:4 — 19) 100 - 150 mm thick laid on blinded hardcore bed (meas. sep.)

Cube × 0.15 = 2.87 m³

SMM F.4.2, F.4.6, F.6.1, F.6.8, F.5.2
Concrete bed measured in m³ thickness classified as F.5.2. Provision for cubing made under description.

&

Tamp surface of unset concrete

SMM F.9
Measured in m² stating type of treatment.

Damp proof membrane.

lgth. of bldg.	5.490
add ²/50	0.100
	5.590

9.

Waste calculation to find size of damp-proof membrane. In this case membrane turns vertically up the face of brickwork 150 mm high and then under the d.p.c. for which 50 mm has been allowed.

			3.490
			0.100
			3.590

5.59			1000 gauge poly.
3.59	20.07	damp proof membrane laid horizontally on blinded hardcore bed with minimum 200 mm laps. (meas. net no all. for laps)

SMM G.37.1, G.37.2
Measured in m².

vertical girth

2/5.490	10.980
2/3.490	6.980
	17.960

| 17.96 | 17.96 | Ditto but vertical to face of brickwork 150 mm deep. |

Damp proof course.

| 2/ 18.98 | 37.96 | 102.5 mm wide bitumen felt hessian based d.p.c. to B.S.743 3mm thick in single horizontal layer with min. 100 mm laps & bedded in c.m.(1:4) (meas. net no allow for laps) |

SMM G.37
Damp-proof courses over 225 mm wide measured in m². Under 225 mm wide measured in linear metres stating width.

10.

Item		Allow for keeping the surface of the site and the excavation free of surface water	*SMM D.25* An item is to be given for keeping the surface of the site and the excavation free of surface water.
		Superstructure.	
		The following brickwork and blockwork in load-bearing superstructure.	*SMM G.3* Brickwork is to be classified as in: (a) foundation; (b) load-bearing superstructure; (c) non-loadbearing superstructure.
18·98 2·40	37·96	Form 50 mm wide cavity in hollow wall a.b.d.	*SMM G.9* Where a description is identical to a previous one in the same take-off a.b.d. means as before described. This saves writing a similar description in full again.
19·59 2·40	47·02	Half B. th. wall in S.O.H.W. ent. in facing bricks Wealden stocks a.b.d. but in g.m. (1:1:6)	*SMM G.14.9* Again similar to the facing brickwork measured below d.p.c. level but gauged mortar specified for super-structure.
		Blockwork.	
		centre line length	. Waste calculation to find centre-line length of blockwork.

11 .

Bwk.

perimeter 20·000

Ddt. corners

4/2/202.5   1·620

18·380

Build up:

external skin	102.5
cavity	50
half width internal skin	50
deduction per corner	2/202.5

18·38
2·40 | 44·11

100 mm thick solid Celcon blockwork in S.O.H.W. size 440 × 215 mm keyed for plaster and bedded & jointed in g.m. (1:1:6)

SMM G.26.1, G.27.1.c
Blockwork to be measured in m^2 stating details listed in G.26.1.
Note classification required in G.27.1 is in skins of hollow walls.

Timber flat roof

no. of joists.

lgth. of bldg. 6·000
Ddt.
2/025   0·050
      400)5·950
= 14·87 = 15+1 = 16

To calculate number of joists find centre to centre dimensions of end joists. Divide by spacing 400 mm centres. To maintain centres at 400 mm 15 joists are required plus 1 for end.

12.

Carcassing timber

16/	4.00	64.00

50×200 mm sawn softwood in flat roof.
(roof joists)

SMM N.2
Measured in metres stating cross-sectional size. Classified as SMM N.2.1. The words in brackets (roof joists) are called a sign post and are to indicate taker of logic and are not taken to bill.
SMM heading.

First fixings

16/	4.00	64.00

50mm sawn softwood firring pieces average 38mm thick.

SMM N.6.1
Measured in linear metres stating width and average depth. Firrings taper 75 to 0 so average is 38 mm.

```
              4.000
fascia width  0.018
              4.018
```

Chipboard passes over fascia board on one side of roof hence waste calculation.

	6.00	
	4.02	24.12

18mm thick butt jointed roofing grade chipboard fixed to softwood bearers in flat roof.

SMM N.4.1, N.4.1.d
Boarding and flooring measured in m² stating thickness and method of jointing and fixing. Classified as N.4.1.d.
Note a flat roof is considered as a roof not exceeding 10° pitch from the horizontal.

	6.00	6.00
2/	4.02	8.04

38×38 mm sawn softwood tilting fillet.

SMM N.10.1
Measured in linear metres stating cross-section dimensions. Tilting fillet is required to 3 sides of roof but not rainwater discharge side.

	14.04

13.

Fascia board.

Joists	225
Firrings	75
Chipboard	18
Tilr fillet	38
Blk. cover	25
	381
Fascia hght. say	385

Waste calculation to find height of fascia board required. This was given in the specification but shows how it can be calculated.

length	6·000
Add mitres	
2/018	0·036
	6·036

Calculation to find lengths of fascia to allow for cutting of mitres.

length	4·000
mitres 2/018	0·036
	4·036

$$2/ \quad \frac{6·04}{4·04} \quad \frac{6·04}{8·08}$$

18 × 385 mm
W.B.P. plywood
fascia board.

SMM N.41.g
Measured in linear metres stating cross-sectional dimensions. N.1.4 states that fixing is deemed to be by nails. Other types of fixing for example by screws would have to be stated. This is the deeper fascia to three sides of the roof.

&

K.P.S. + ②
to ditto new
work externally
Super x 0.403 = 5.69m²

SMM V.1.2, V.3.1, V.3.2, V.4.1.g
It is convenient to add painting to fascia at this stage. The paint surface in this case is 385 for the front edge plus 18 for the bottom edge which is over 300 mm girth and is measured in m². Note the provision for squaring under the description.

&

Prime only back
of fascia board
prior to fixing
new work extly.
Super x 0.403
= 5.69 m²

The primed area is the back edge plus top edge = 385 + 18 = 0.403. Painting should be classified as V.1.2.
New work internally
New work externally
Repainting and redecoration internally
Repainting and redecoration externally
In the bill the painting to fascia is classified as V.4.1 to general surfaces.

| 14.12 | 14. |

		joist depth 225
		brick cover 25
		250

<table>
<tr><td>6.04</td><td>6.04</td><td>18 × 250 mm
W.B.P. plywood
fascia board.

&

K.P.S. ⨁ ②
to ditto 150 –
300 mm girth
new work extly.

&

Prime only back
of fascia bd.
prior to fixing
150 - 300 mm
girth new work
extly.</td></tr>
</table>

Waste calculation to calculate height of smaller fascia board on rainwater discharge side of roof.

The painted area to the previous fascia board exceeded 300 mm girth and was therefore measured in m² clause V.3.2. In this case the girth is:

front face	250
bottom edge	18
	268

This falls within the girth range of 150–300 and is measured in linear metres. The prime only area is:

back face	250
top edge	18
	268

and is also within the 150–300 girth range and therefore measured in linear metres.

4		Mitres to fascia board.

SMM N.1.10
Fascia boards cross-sectional area exceeds 0.002 m² and mitres at corners of building are therefore enumerated.

6.04		25 × 38 mm sawn softwood drip.

SMM N.6.2
Drips given in linear metres stating cross-sectional dimensions.

end of first fixings

15.

Item	Allow for protection to excavation and earthwork.	Protection items
		SMM D.45
	&	
	Ditto concrete wk.	SMM F.45
	&	
	Do. bwk. & block.	SMM G.58
	&	
	Do. woodwork	SMM N.33.2
	&	
	Do. painting & decorating.	SMM N.13

Bill of quantities for Figs. 8.29, 8.30 and 8.31

EXCAVATION AND EARTHWORK

Preamble clauses would normally precede measured work.

Plant

A   Allow for bringing to and removing from site all plant necessary for this section of the work.   item

B   Allow for maintaining on site all plant required for this section of the work.   item

Site preparation

C   Excavate topsoil to be preserved average 150 mm deep.   28   m^2

Excavation

D   Excavate foundation trench exceeding 0.30 m wide commencing at surface strip level maximum depth not exceeding 1.00 m.   9   m^3

Earthwork support

E   Earthwork support maximum depth not exceeding 1.00 m and width between opposing faces not exceeding 2.00 m.   32   m^2

Disposal of water

F   Allow for keeping the surface of the site and the excavation free from surface water.   item

Disposal of excavated material

G   Remove surplus excavated material from site.   9   m^3

H   Deposit preserved topsoil in temporary spoil heap for re-use average 25 m from excavation.   4   m^3

Filling

I   Earth filling around foundations with topsoil from permanent spoil heap.   1   m^3

J   Hardcore filling obtained off site in making up levels average 150 mm thick.   19   m^2

Surface treatments

K   Level and compact base of excavation to receive concrete.   11   m^2

L   Level and compact surface of ground to receive hardcore filling.   16   m^2

Blinding

M   Blind surface of hardcore with sand.   19   m^2

Protection

N   Allow for protection to all work in this section.   item

To summary   £

1.

CONCRETE WORK

Preamble clauses and plant items.

The following in OTHER CONCRETE work

approximate total volume 12 m³.

	In-situ concrete		
A	Concrete (1 : 3 : 6 – 19) in foundation trenches over 300 mm thick poured against face of excavation.	9	m³
B	Concrete bed (1 : 2 : 4 – 19) 100–150 mm thick laid on blinded hardcore bed.	3	m³
C	Weak concrete (1 : 12) filling to hollow wall not exceeding 100 mm thick and under 1 m³ in size.	1	nr.
D	Tamp surface of unset concrete.	19	m²
	Protection		
E	Allow for protection to all work in this section.	item	

To summary   £

2.

BRICKWORK AND BLOCKWORK

Preamble clauses and plant items.

Foundations

Brickwork

A   Half brick thick wall in skins of hollow walls in calcium silicate bricks class 3 in stretcher bond in cement mortar (1 : 4).   7   m²

B   Form 50 mm wide cavity in hollow wall including 4 number galvanized butterfly wall ties to B.S. 1243 per m².   6   m²

Brick facework

C   Half brick thick wall in skins of hollow wall entirely in facing bricks (Wealdon stocks) in stretcher bond in cement mortar (1 : 4) pointed with a neat rubbed-in joint.   5   m²

Damp-proof courses

D   102.5 mm wide bituminous felt hessian based damp-proof courses to B.S. 743 in single horizontal layer with minimum 100 mm laps and bedded in cement mortar (1 : 4) measured net, no allowance for laps.   38   m

E   1000 gauge polythene damp-proof membrane laid horizontally on blinded hardcore bed with minimum 200 mm (measured net, no allowance for laps.)   20   m²

F   Ditto to vertical face of brickwork 150 mm deep.   18   m

Load bearing superstructure

Brickwork

G   Form 50 mm wide cavity in hollow wall including 4 number galvanized butterfly wall ties to B.S. 1243 per m².   46   m²

Brick facework

H   Half brick thick wall in skins of hollow wall entirely in facing bricks (Wealdon stocks) in stretcher bond in gauged mortar (1 : 1 : 6) pointed with a neat rubbed-in joint.   47   m²

Blockwork

I   100 mm thick solid Celcon blockwork in walls and partitions size 440 × 215 mm keyed for plaster bedded and jointed in gauged mortar (1 : 1 : 6)   44   m²

Protection

J   Allow for protection to all work in this section.   item

To summary   £

3.

WOODWORK

Preamble clauses.

Carcassing timber

Impregnated sawn softwood as described

A	50 × 225 mm in flat roof.	64	m

First fixings

Boardings and flooring

B	18 mm thick butt jointed roofing grade chipboard fixed to softwood bearers in flat roof.	24	m
C	18 × 250 mm W.B.P. plywood fascia board.	6	m
D	18 × 385 mm ditto	14	m
E	Mitres to fascia board.	4	nr.

Firrings drips and bearers

Impregnated sawn softwood as described

F	50 mm wide firring pieces average depth 38 mm	64	m
G	25 × 38 mm drip.	6	m

Fillets and rolls

Impregnated sawn softwood as described

H	38 × 38 mm tilting fillet.	14	m

Protection

I	Allow for protection to all work in this section.	item

To summary   £

4.

PAINTING AND DECORATING

Preamble clauses.

New work externally

Painting polishing and similar work

Prime only back of woodwork prior to fixing

A	General surfaces.	6	m²
B	Ditto on isolated surfaces 150–300 mm girth.	6	m

Knot, prime, stop and paint two undercoats and two coats of oil paint to woodwork

C	General surfaces.	6	m²
D	Ditto on isolated surfaces 150–300 mm girth.	6	m

Protection

E	Allow for protection to all work in this section.	item	

To summary   £

5.

Page nr.	SUMMARY			£	
1	Excavation and earthwork				
2	Concrete work				
3	Brickwork and blockwork				
4	Woodwork				
5	Painting and decorating				
	FINAL TOTAL CARRIED TO FORM OF TENDER			£	

Note:

The purpose of the summary sheet found at the rear of the bill of quantities is to collate the collection totals from each work section. The final total is obtained by adding together all of the collection figures and this amount is inserted on to the form of tender as the contractor's tender sum.

In practice a summary would be more detailed than this simple example, depending on the size and complexity of the project. Students are advised to compare several bills of quantities to form an idea of the contents and layout of a summary.

6.

Chapter Nine

Measurement on site

Measures executed work under working conditions

So far we have considered measurement on a pre-contract basis, that is, for the preparation of contract documents prior to the commencement of work on site. Once work has begun there will almost certainly be a necessity for work to be measured on site. Let us consider some of the reasons why and for what purposes such measurement is required, firstly from the quantity surveyor's point of view and then the contractor's.

Measurement on site by quantity surveyor

1. *Provisional measurements in bills of quantities*
It is common practice in bills of quantities for certain work to be measured provisionally. Examples of these would be excavation, drainage and external work. The word 'provisional' used in this instance means that the bill of quantities has been prepared as accurately as possible, but site conditions may alter assumptions made at the time of preparing the bill of quantities and therefore the listed quantities. Should this happen the work is measured on site and the contractor paid in accordance with the work actually carried out at the prices inserted by him in the rate column of the bill of quantities.

2. *Variations and alterations*
Where variations and alterations occur it is often necessary physically to measure the work on site. It happens occasionally that work becomes so varied that it is necessary to measure the whole work again during the progress of the contract.

3. *Valuations*
In preparing interim valuations and making assessments of work in progress it may be necessary to take measurements to value the proportion of work complete.

4. *Existing buildings*
On existing buildings it may be necessary for the quantity surveyor to measure

Measurement by the contractor

The contractor's surveyor would also undertake similar measurement to that previously outlined by the quantity surveyor, to ensure that the contractor is being correctly paid for the work carried out. Additionally for the following purposes:

1. to calculate quantities of materials for ordering purposes;
2. many building operatives are paid bonus based on output. For bonus to be calculated the work has to be measured. Larger companies retain staff known as measuring surveyors specifically for this purpose;
3. to assess work carried out by labour-only sub-contractors;
4. for use in updating contract programmes;
5. on works to existing premises to enable the contractor to assess the quantity of work to be carried out.

Measurement tools

These are few and may be listed as follows:

1. tape;
2. measuring rod or rule;
3. dimension paper or dimension book.

Tapes
These can be obtained in many lengths in cotton or steel. In the main they are more useful for measuring larger distances.

Measuring rods
The folding rigid wooden 2 m measuring rods are particularly suitable for awkward areas, inside buildings and for widths and depths of trenches – see Fig. 9.1.

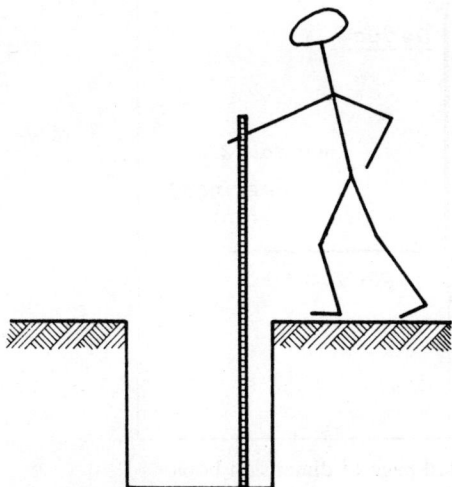

Fig. 9.1 Measuring depth of excavation

On site there is no order of measurement as such. The surveyor should consider what measurements are necessary and use a logical order. The practice of measuring full and then deducting for voids or wants makes for consistency.

The dimensions may be recorded on sheets of dimension paper attached to a clip board or alternatively in a dimension book. Dimension books are hard bound and have the advantage that the pages are not loose and not so easily lost. When the book is full it may be easily filed so that there is easy reference should the need occur. A surveyor often needs to spend a whole week on various sites taking measurements and then comes back to the virtual peace of the office to square up and calculate the dimensions recorded.

Typical page of dimension book
A dimension book is ruled up in a similar manner to dimension paper, but has only one set of columns. Each page is pre-numbered. The following example illustrates a page of a dimension book for re-pointing work.

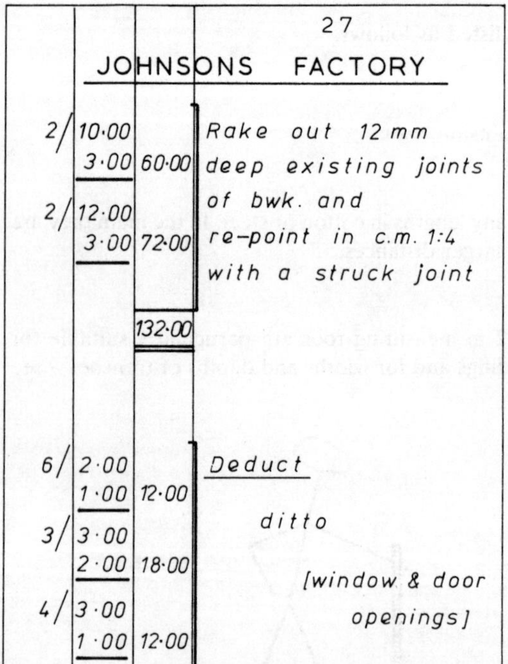

Fig. 9.2 Completed page of dimension book

Part Three

Estimating

Chapter 10

Estimating

Introduction

The purpose of this section of the book is to introduce students to estimating in general terms only. To estimate successfully for building works calls for many years of experience. At this level students should seek to understand general principles, where to find information and how to apply it.

Approximate estimating using published data sources

Information on building costs are available from several sources. Most are easily accessible to students and are listed here for guidance.

Price reference books
Well-known titles include:
1. Spon's *Architects and Builders Price Book*;
2. Laxton's *Building Price Book*;
3. Griffith's *Building Price Book*;
4. Hutchins *Priced Schedules*.

   Price books are published annually at the beginning of each new year. As prices increase through the course of a year the information gradually becomes outdated. Estimators are usually aware of percentage increases, however, and can make their own adjustments to book prices.

   Layout and presentation of the individual books varies, but the following information is often included: calculation of basic wage rates; daywork rates; professional fees for architects, quantity surveyors and consultant engineers; market prices for building materials; labour constants; prices for measured work; approximate estimating sections.

   Of most interest to students will be the sections dealing with prices for measured work. These are laid out in the same order as the SMM and prepared bills of quantities. By using the rates given students will be able to calculate approximate prices for building work for comparative purposes and also for use in project exercises. Example 10.1 is taken from the concrete work section of a typical price book and shows how to use the information.

Example 10.1

Plain concrete

		Concrete mixes		
		1:3:6	1:2:4	1:1½:3
		£	£	£
In foundation and column bases.				
Over 300 mm thick	m³	29.85	33.90	37.95
150–300 mm thick	m³	30.66	34.71	38.75
100–150 mm thick	m³	31.47	35.52	39.56
Not exceeding 100 mm thick	m³	32.28	36.33	40.38

To find the rate, then, for plain concrete (1:2:4 – 20) exceeding 300 mm thick in foundations place a ruler under the relevant description and read off under the 1:2:4 mix column which in this case would be £33.90 per m³. The price of £33.90 will have been built up by analysing the work under the following headings:

1. material content;
2. labour content;
3. plant content;
4. addition for overheads and profit.

The introduction to the price book should be carefully studied to determine the basis used to calculate market prices, labour rates and what percentage has been added to cover overheads and profit for the rates given.

Periodical magazines

In addition to price books, many periodicals concerned with the construction industry contain frequent supplements on building costs. Popular titles include:

1. *Building*;
2. *Building Trades Journal*;
3. *Q.S. Weekly*.

Whatever the source of published information used it should be understood that the prices given are an average for guidance only. There is no substitute for an estimator calculating a rate for work based on known facts relating to a specific project. The size of a building company, its efficiency, workload, labour force, standard of workmanship and level of profit required will all affect the estimator's calculation.

Components of unit rates and estimator's procedure in item pricing

Unit rates

The estimator calculates a unit rate for carrying out a specific item of building work in terms of cost per linear, square or cubic metre, or number of, and in the case of reinforcement and structural steelwork in tonnes. The price is then inserted in the rate column of the bill of quantities and timesed by the given quantity of the work to arrive at the total cost for that operation.

The estimator calculates a unit rate by breaking any operation into the same elements that were described in price books, namely materials, labour and plant.

Materials
Basic material costs should be obtained from suppliers and allowances added for waste and, depending on the type of material, unloading time.

Labour
Here the estimator must use his judgement and experience to calculate labour content. He may be assisted by labour constants published in price books and estimating supplements. These give the time in hours or parts thereof that it should take a craftsman or labourer to carry out a specific operation. These must be viewed with caution as they are only averages.

It should be added that many contractors employ labour-only sub-contractors who agree to undertake work for a certain price per linear, square or cubic metre. This simplifies the estimator's task to a certain degree.

The calculation of the labour element is, however, the most difficult part of calculating a unit rate.

Plant
Plant is included in certain unit rates, but not in others. Some estimators, for instance, prefer to think in terms of how many weeks a mixer will be required and site and price it accordingly on a lump sum basis rather than include its price into, say, the unit rate for brickwork in terms of price per m^2.

The total of materials, labour and plant for a unit rate is known as the **net** cost. When an addition for overheads and profit is made it is known as the **gross** cost.

Examples 10.2 and 10.3 illustrate the calculation of unit rates.

Example 10.2
The following description appears in a bill of quantities:

		Quan.	Unit	Rate	£	p
B	Excavate foundation trench exceeding 0.30 m wide commencing at surface strip level, maximum depth not exceeding 1.00 m	150	m^3			

The estimator must calculate a rate for this item. For convenience this has been done on double column cash paper. (*See* Unit rate for Example 10.2 on p. 198.)

Example 10.3

		Quan.	Unit	Rate	£	p
C	One brick thick wall in plain commons in Flemish bond in gauged mortar (1 : 1 : 6)	100	m^2			

Unit rate for Example 10.2	£	p	£	p
MATERIALS				
None				
LABOUR				
Labourer acting as banksman and assisting machine at hourly all-in wage rate of £3.30 per hour	3	30		
Total cost of labour per hour	3	30	3	30
PLANT				
Assume rubber-wheeled backacter excavator. Hourly hire rate including driver and fuel, say £8.00 per hour	8	00		
Total cost of plant per hour	8	00	8	00
∴ Total hourly cost			11	30

It is at this point that the estimator must use
his experience and judgement to evaluate
the likely output of the machine per hour.
Output will depend on:
1. operator's ability and motivation;
2. complexity of the work;
3. type of ground conditions.
Assume that the estimator decides on an
output of 6 m³ per hour,

$$\therefore \text{cost per m}^3 = \frac{\text{hourly cost}}{\text{output}}$$

$$= \frac{£11.30}{6} = £1.88$$

This represents a net cost to the contractor
of £1.88 per m³ and to this figure must be
added an addition for overheads and profit.

To calculate the unit rate for Example 10.3, the estimator is assumed to be working on the following information:

cost of bricks £40 per 1000
mechanical unloading of bricks £1.50 per 1000
waste on bricks 5%
mortar £22 per m³ (0.07m³ of mortar required per m² of brickwork)
waste on mortar 10%, and

brickwork gang 2 bricklayers and 1 labourer. Hourly cost of gang £11. Output per hour for gang, say, 120 bricks.

	£	p	£	p
Unit rate for Example 10.3				
MATERIALS				
Basic cost of 1000 bricks	40	00		
Lorry-mounted crane unloading per 1000 bricks	1	50		
Cost of 1000 bricks on site	41	50		
Cost per m² required				
118 bricks per m² in one brick thick wall				
$\therefore$ cost per m² $= \dfrac{\text{cost per 1000}}{1000} \times$ number of bricks required per m²				
$= \dfrac{\text{£41.500}}{1000} \times 118 = \text{£4.897}$	4	90		
Waste on bricks 5%	0	25		
Mortar 0.07 m³ per m² at £22 per m³ = 0.07 × £22	1	54		
Waste on mortar 10%	0	15		
Total cost of materials per m²	6	84	6	84
LABOUR				
Cost per m² $= \dfrac{\text{hourly gang cost}}{\text{output}} \times$ number of bricks required per m²				
$= \dfrac{\text{£11}}{120} \times 118 = \text{£10.816}$ $\dfrac{\text{£11}}{120}$ gives cost of laying one brick × 118 gives cost per m²	10	82		
Total cost of labour per m²	10	82	10	82
PLANT				
The cost of the mixer assumed calculated in with the mortar cost.				
Net cost of brickwork per m²			17	66
Gross cost add for overheads and profit 15%			2	65
Total cost per m²			20	31

Overheads and profit and all-in labour costs

When calculating unit rates and in looking at price books, mention was made of both these terms. What exactly is meant, then, by overheads and profit and all-in labour costs?

Overheads and profit

Overheads for a building company is the cost of head office administration or, looking at it in another way, the cost of running the organization. Consider some typical costs of running the business:

Office accommodation – rent, rates, equipment, furnishings, maintenance, insurance;

office lighting and heating;

telephone and postal charges;

salaries;

company vehicles – petrol, tax, insurance, servicing, replacement, etc.;

stationery and incidentals.

The company must endeavour to recover the cost of overheads from the projects it tenders for. To do this it must calculate the cost of overhead expenses for the year and predict the likely value of work to be obtained.

Example 10.4. Our theoretical company establishes from the previous year's records that it expects its head office overheads to be £200,000 in the current year. It anticipates carrying out £2 million worth of work through the year:

estimated cost of overheads for year = £200,000

anticipated value of work to be undertaken = £2,000,000

$$= \frac{£200,000}{£2,000,000} \times \frac{100}{1} = 10\%$$

To recover the cost of overheads then 10 per cent must be added to each project tendered for.

Profit

Building companies expect to earn a profit when they carry out work. The building industry covers a wide variety of work and profit levels vary greatly within the various sectors of the industry. Companies are also wary of disclosing profit and overhead margins which may be of use to competitors. Again, levels of profit will depend on the state of the economy and the amount of building work available at any given time.

In general, though, profit margins obtained in the tender system by contractors may be as low as $2\frac{1}{2}$ per cent for large works up to perhaps 25 per cent plus for minor works.

To a net unit rate or contract tender figure, then, the contractor will add a percentage addition to cover overheads and profit. In many text-books the student will see 15 per cent added to cover these costs. Not too much notice of the actual percentage figures should be taken, but rather of the principles behind its addition.

All-in cost of labour

Most building operatives employed by contractors are paid on a weekly basis. For convenience in estimating this can be expressed as an hourly rate by dividing the weekly wage by the number of hours worked in a normal week. This hourly wage is known as the operatives' **basic rate.** There is a nationally agreed minimum weekly wage for craftsmen and labourers, but many contractors pay above that level in the form of bonus payments or higher basic rates.

The actual cost of emplying an operative is far higher than the basic rate that he receives. This is because the contractor must pay national insurance, contribute to holiday pay and Construction Industry Training Board (CITB)

levies, etc. in respect of each operative. It is necessary for estimating purposes
for the contractor to establish the actual cost per hour to employ a person and
this is known as the **all-in rate**. To find the all-in hourly cost means calculating
the total yearly or weekly cost of that person and then dividing that figure by
the number of hours worked in a year or week. The building contractor has to
be able to calculate this figure because this is the rate he must charge per hour
for that operative to cover his costs before he begins to make any profit.

Rather than give an example in this book which would quickly become out
of date the student is referred to the *Code of Estimating Practice* published by
the Chartered Institute of Building. For a worked example of current all-in
rates the student should refer to the estimating supplement of the *Building
Trades Journal*.

Discount associated with building work

The estimator must be familiar with the level of discounts obtained by his
company. There are various types of discount and these are listed below:

1. trade discounts;
2. cash discounts;
3. discounts to main contractors by nominated sub-contractors and suppliers.

Trade discounts
Most suppliers of building materials allow contractors a discount when
purchasing goods from them. This is called a trade discount. The level of
discount will depend on the quantities being purchased, the general level of
business between the supplier and contractor and the contractor's reputation
or lack of it with respect to prompt payment for the goods. Trade discounts can
vary between 10 and 40 per cent, depending on the type of material. A
contractor should ensure that he has a written quotation stating the basic cost
of the material and the trade discount applicable. Vague verbal quotations of
say 20–25 per cent off list price should be avoided.

Cash discounts
This title is a little misleading. Most contractors have credit accounts with
material suppliers. In effect this means that goods purchased during the
current month are not paid for until at the latest the end of the following
month. To encourage prompt payment of accounts some suppliers offer a cash
discount of $2\frac{1}{2}$ per cent on the cost of the goods supplied provided they are
paid for before the last day of the month following the month of supply.

For example a contractor purchasing goods on 1 or 2 July would not need
to make payment until 31 August to claim the discount. This amounts to
virtually two months of credit for the contractor.

Nominated suppliers and sub-contractors' discounts
Under the JCT form of contract the main contractor is entitled to the following
discounts in respect of nominated sub-contractors and suppliers:
nominated sub-contractors – $2\frac{1}{2}$ per cent of the value of the
nominated sub-contractor's work;
nominated suppliers – 5 per cent of the value of the goods supplied by a
nominated supplier.

Value added tax (VAT)

Students will most probably already be familiar with this form of taxation which is in use in all European Economic Community (EEC) countries. A certain percentage (currently 15%) in the United Kingdom is added to the cost of many goods and services. Value added tax is administered by H.M. Customs and Excise and all companies exceeding a specified annual turnover are required to be registered with them. Those that do not come up to the minimum annual turnover may apply to be registered if they so wish, but do not have to do so.

We are concerned here with the effect of VAT on the construction industry and in particular on the estimating process.

New construction work
New construction work is zero rated. This means that the contractor does pay VAT on invoices rendered for goods and services supplied, but he is able to claim the whole amount back from H.M. Customs and Excise. The estimator in this case need not concern himself with VAT in the calculation of the estimate. The company may, however, allow in its overheads cost a certain percentage to cover the cost of administering the scheme.

Work of a repair, renewal or redecoration nature
This work is not zero rated and the contractor must pass on to the client any liability for VAT. Much construction work is a combination of new and repair work and this causes the most problems in the implementation of VAT; H.M. Customs and Excise will always give advice where necessary.

Forms of contract and VAT
Most forms of contract associated with building work make provision for the contractor to be reimbursed from the client for any VAT payments related to the works which are not reclaimable by the contractor.

Factors affecting cost

We have seen how an estimator goes about building up unit rates and how overheads and profit are dealt with. Not all work is of a similar nature, however, and therefore each project must be considered on its own merits and unit rates worked out accordingly. That is why rates in price books can only give a general guide, they cannot possibly relate to all types of work.

A major factor affecting cost is the general state of the economy. It is a fact of life that when governments seek to save money a favourite area is the public building sector. Cuts are made in projects such as housing, public buildings, roads and other construction work. The result is that competition for work becomes fierce and profit margins are reduced. At other times money is spent by governments on building schemes in an attempt to reduce unemployment and increase investment. During such periods there is often a surplus of work and margins and therefore costs are increased accordingly.

Apart from the general state of the economy other factors which affect cost include scale, repetition and time.

Scale
Generally speaking, the larger the project then the less it will cost in *pro-rata*

terms. Overheads will be less than with smaller projects. Because of the volume of work it is more efficiently organized and labour and plant can be fully utilized. For example, on a small site a machine may be required for only two hours. Most plant is charged on a minimum basis of a day or week's hire and therefore in this case it would be inefficiently used and costly. Compare this to a larger site where plant may be hired in for, say, two months at advantageous rates and one can quickly see how scale affects costs.

In practice the economics of scale are reduced somewhat by specialization. Most national contractors are only interested in schemes of a certain minimum value and above, and are content to leave the medium-range contractors to deal with middle-sized local schemes. This specialization carries right down to the one-man business with minimum overheads which may carry out repair work and the like.

Repetition

If a factory producing motor cars is visited one will find a highly organized procedure with the finished product rolling off the production line every few minutes. This does not happen by chance, but rather because improvements have constantly been made to a repetitive process to make it as efficient as possible.

When a building is constructed it is usually a one-off project and decisions as to the best methods of construction have to be made. Often when work is finished it is apparent that other methods would have been better and could have reduced costs.

Certain building work is of a repetitive nature. For instance, large housing schemes where the designs are of an identical pattern. Here the contractor can expect to improve efficiency and productivity as the scheme progresses and the best construction methods are found, enabling savings in costs to be made which are normally reflected in the tender prices submitted initially.

Time

The length of a project in terms of months and the time of year at which building work is to be undertaken can also greatly affect costs.

Building contractors constantly have to seek new work to keep their staff and labour force employed. Longer projects lasting two years and over offer security and constant work over a long period and are fiercely competed for. Building projects that are short term and required quickly often have penalty clauses against the contractor should he fail to complete on time. Such projects tend to attract higher tenders because of the risks involved.

The time of the year during which a project is to be built can also affect costs. Winter projects often place estimators in a dilemma. On the one hand most contractors seek work for the winter months, when work is scarce, to keep the labour force together. Against this are the risks of winter building, particularly work in the ground and shell of the building where bad weather can play havoc.

Conclusion

The student should now realize that although unit rates may be built up in a logical manner there are many other variables which can affect costs and tender prices generally, and therefore how important experience of the industry as a whole is in estimating.

Tendering processes

This is the manner in which contractors will be invited to submit tenders for work. The system used is at the discretion of the architect for private works or the local authority or government department for public works.

The common systems of tendering are explained below and may be listed as follows:

1. open tendering;
2. selective tendering;
3. negotiated contracts.

Open tendering

With this system details of a project are given in the local, national and/or trade press. Tender documents are sent to contractors who express an interest in the project. The contractor submitting the lowest tender figure will normally be awarded the contract, although the lowest price does not have to be accepted.

Because of their public accountability local authorities often use this system. The disadvantages of open tendering are fairly obvious but may usefully be stated:

1. When work is scarce many contractors may apply to tender for the work. Contract documents are not cheap to prepare and each contractor will require a set. Only one contractor can be successful and much time and effort will be wasted. The unsuccessful contractors must try and recover the cost of tendering against other projects.
2. The contractor submitting the lowest tender may not have the ability, resources or experience required for the project and this may lead to problems during the course of the work.
3. There is always a lengthy delay to a contract being awarded due to time taken in advertising the project, receiving replies, forwarding tendering documents and finally receiving tenders.

On the plus side the system does allow new firms to obtain work which might otherwise be denied to them and to prove their capabilities.

Selective tendering

This differs from open tendering in that selected contractors are invited to tender for the project. This may be done by architects or local authorities writing to companies who have the experience and known ability to carry out the project. Alternatively, some architects and local authorities maintain lists of approved contractors who have carried out works satisfactorily in the past. The inclusion of a contractor's name on such a list may be of great value to that firm in ensuring a steady flow of enquiries and may encourage the company to maintain its standards.

As with the open tendering system the lowest tender price is normally accepted.

Useful guidance *for selective tendering* is given in the *Code of Procedure* produced by the joint Consultative Committee for Building.

Negotiated tenders

Negotiated tenders can take many forms, but they basically occur when a contractor is selected to carry out a particular contract by the architect or client. There can be several reasons for this for example:

1. specialized knowledge of the contractor;
2. perhaps the contractor has carried out previous contracts for the client or architect which have been satisfactory;
3. the building is required as quickly as possible and by negotiating a contract the delays which are normal with the open and selective systems are reduced.

Several major contractors now actively seek negotiated contracts claiming that the knowledge and expertise that they offer can actually save the client money if they are called in at an early stage.

Costs and values

It is perhaps appropriate that at the end of this book we should look at the significance of cost and value to the contractor and the client. After all a completed building, alteration or repair is what the construction industry is all about. Measurement is just part of that process. Ideally the contractor should have made a profit from the work and be able to take some pride in the finished product. The client should be pleased with the work and feel that he has achieved good value for money. Where the building is for the public at large hospitals, schools, libraries, etc. then the community as a whole should feel pride in the building and it should enhance the value of life to the public as a whole.

Significance of cost and value to the contractor

Cost
The cost of undertaking building work to the contractor is what has to be paid out in total on material purchases, labour and plant charges, together with overhead costs.

Value
Value to the contractor is initially in the profit that is made from undertaking the work. This can be defined as:
Total final contract figure − cost = profit
 Other factors of value to the contractor are:
1. further work and recognition which may result from carrying out specific projects;
2. enables the labour force of a contractor to be kept together;
3. regular monthly payments enables contractor to meet bills and overhead expenses and to maintain a steady cash flow which is very important.

Significance of cost and value to the client

Cost
Cost to the client of a building project may include the following:
Building cost + professional fees + cost of land + interest charges on borrowed capital
It may be of course that the client already owns the land. Indeed, the client may be able to finance the project himself, but then that money could have been invested or used elsewhere so there is always some financial charge involved in building work.

Value to a client is harder to define because it will depend on the type of building and its use. Let us look at a few examples:

1. To a **householder** an extension to a home may provide more living space, improved quality of life and the increase in value of the property will often exceed the cost of the work.

2. To a **factory owner** a new factory or alteration may increase a company's output and profit and provide better working facilities for the labour force. The increased value of the building will improve the asset value of the company and can therefore, if required, enable additional finance to be borrowed to allow for greater expansion.

3. A new library, school or hospital, properly designed and built, should give many years of valuable service to a community. Good-quality, interesting buildings tend to improve a town or city's image, and apart from their functional uses may encourage tourism and therefore trade and commerce to an area.

4. To a **developer** the value of a building is the profit he hopes to make when he sells a completed building to an owner. The owner may let the building to other occupiers who will trade from the premises. In cases such as this there may be value to many parties in the erection of one new building.

Further reading list

'Careers in Building': leaflets available from the Chartered Institute of Building.
'Careers in Quantity Surveying': leaflets available from the Institute of Quantity Surveyors.

Related books
Construction Technology Vols 1–4, by R. Chudley. Longman.
Building Services and Equipment, Vols 1–3 by F. Hall. Longman.
Building Organisation and Procedures, by G. Forster. Longman.

Titles on quantity surveying
Building Quantities Explained, by I. H. Seeley. Macmillan.
Elements of Quantity Surveying, by A. J. Willis and C. J. Willis. Granada Publishing.
Measurement of Building Work, by W. H. Wainwright and R. J. Whitrod. Hutchinson.

Estimating
Code of Estimating Practice. Chartered Institute of Building.
Practical Builders Estimating, by W. H. Wainwright and A. A. B. Wood. Hutchinson.

Index